Heart *of the* Jaguar

ALSO BY JAMES CAMPBELL

*Braving It: A Father, a Daughter, and an
Unforgettable Journey into the Alaskan Wild*

*The Color of War: How One Battle Broke
Japan and Another Changed America*

*The Ghost Mountain Boys: Their Epic March
and the Terrifying Battle for New Guinea—
The Forgotten War of the South Pacific*

*The Final Frontiersman: Heimo Korth and His
Family, Alone in Alaska's Arctic Wilderness*

HEART
of the
JAGUAR

The Extraordinary Conservation
Effort to Save the Americas'
Legendary Cat

JAMES CAMPBELL

W. W. Norton & Company
Independent Publishers Since 1923

Copyright © 2025 by James Campbell

All rights reserved
Printed in the United States of America
First Edition

For information about permission to reproduce selections from this book, write to
Permissions, W. W. Norton & Company, Inc., 500 Fifth Avenue, New York, NY 10110

For information about special discounts for bulk purchases, please contact
W. W. Norton Special Sales at specialsales@wwnorton.com or 800-233-4830

Manufacturing by Lakeside Book Company
Book design by Beth Steidle
Production manager: Lauren Abbate

ISBN 978-0-393-86761-9

W. W. Norton & Company, Inc., 500 Fifth Avenue, New York, NY 10110
www.wwnorton.com

W. W. Norton & Company Ltd., 15 Carlisle Street, London W1D 3BS

10 9 8 7 6 5 4 3 2 1

*For Elizabeth and our daughter Willa,
who first joined me in jaguar country.*

*And to the memory of Howard Quigley,
whose life continues to inspire so many.*

His stride is wildernesses of freedom.

Ted Hughes, "The Jaguar"

Contents

Preface	XIII
INTRODUCTION: The Rock Star	1
1 The Jaguar Craze	11
2 Bolivian Cocaine	18
3 The Dream	23
4 The God of Death	31
5 The Survivor	38
6 The Hunter	44
7 People of the Jaguar	48
8 The Edge	54
9 He Who Rides a Tiger Cannot Dismount	65
10 *El Tigre*	73
11 The Magic	79
12 The Mother Lode	86
13 A Dream Undone?	95
14 A Bold Conservation Model	102
15 Green Gold	106
16 Land of Thirst	119
17 Field of Dreams	130

18	The Jaguar Corridor	132
19	The Billionaire Who Loved Big Cats	139
20	The Pantanal	148
21	*Onça*	159
22	The Pyrocene	166
23	The Lucky Cats	173
24	Return to the Pantanal	180
25	Touching Wild	188
26	The Magic Cats	197
27	The Tragic Cat	206
28	Land of the Jaguar	216
29	AMLO's Iron Horse	223
30	A Species on the Brink	228
31	A Coming Crisis	233
32	The Beautiful Cat	237
33	Jaguar Interrupted	246
34	Bringing Jaguars Home	253
	EPILOGUE	263

Acknowledgments	269
Notes	273
Sources	287
Index	307

Preface

MY INTEREST IN JAGUARS BEGAN AFTER I READ ALAN Rabinowitz's *Jaguar: One Man's Struggle to Establish the World's First Jaguar Preserve*, the book in which he established his reputation as a scientist, an impassioned conservationist, an adventurer, and a skillful memoirist unafraid of baring his soul. Iron-willed and virtually fearless, he became my new hero.

In 2002, while freelancing for magazines, I proposed to my editor an interview with him. By then he had become something of a celebrated scientist. In August 1999 *The New York Times* wrote, "If Indiana Jones, the daredevil hero of Steven Spielberg's adventure-film trilogy, had been invented as a zoologist rather than an archeologist, Dr. Alan Rabinowitz would have made an ideal model."[1]

Rabinowitz and I spent hours on the phone, talking about his life, his devotion to big cats, and irony of ironies, his allergy to cats, which he didn't discover until his first interactions with jaguars in the Brazilian Pantanal and Belize. After the interview appeared, we discussed the possibility of my joining one of his expeditions to Myanmar and writing a book about him, but he was ambivalent about the idea. He had already penned *Jaguar*, *Chasing the Dragon's Tail* and *Beyond the Last Village*. Over the course of six months, we talked on the phone and emailed. He spoke freely about the extraordinary difficulties of trying to balance his love of field research in far-flung places with his love for his family. He also spoke about his disease, chronic lymphatic leukemia, which doctors diagnosed in 2001 and told him would eventually kill him. Ultimately, he decided to write his own book. That book, *Life in the Valley of Death: The Fight to Save Tigers in a Land of Guns, Gold, and Greed*, about his death-defying experiences in Myanmar, is one of my favorites.

Rabinowitz and I corresponded for a few years, and then we lost touch. In the meantime, his reputation continued to grow. In 2008, *Time* magazine's Bryan Walsh wrote, "Alan Rabinowitz knows tough. The director of the Wildlife Conservation Society's science and exploration program, Rabinowitz made his bones as a young zoologist who would go anywhere to map the shrinking habitats of big animals. He's endured 500-mile hikes through pure jungle, survived malaria, leech attacks, shaky flights on questionable airlines and virtually every other threat that comes from walking the wild parts of the world . . . and even at 53, the muscle-bound Rabinowitz looks like he could wrestle a boa constrictor, and win."[2] In 2010, just a few years after he had left the Wildlife Conservation Society (WCS) for Panthera, an organization devoted to preserving the world's wild cats and the ecosystems they inhabit, I heard an interview with him on *On Being*. "I always thought I could fight everything," he told the host, Krista Tippett. "Just as I was starting to get a bit tired and actually considering slowing down, now I'm told that I have cancer. . . . Now I will never, never wake up a day and sit back and think, 'This is enough.' All I ever do is think of the things that I haven't done yet and that still need doing."[3]

I didn't reconnect with Rabinowitz until the fall of 2017. I had just read an article about an exciting endeavor of his called Journey of the Jaguar. He and his old friend wildlife biologist Howard Quigley were planning to visit countries in Central and South America in hopes of telling firsthand stories about the jaguar. We also talked about his health, and he confessed that he was living on borrowed time.

My relationship with Alan Rabinowitz had begun as hero worship, but when I heard that he had lost his battle with cancer in August 2018, I felt as if I'd lost a friend. I knew, too, that the conservation movement had lost a visionary and a crusader, a man who had devoted—and risked—his life to protect big cats, and especially the jaguar. Jane Goodall said of his death, "Conservation has lost a dedicated, passionate, and powerful voice. On behalf of the big cats, especially the jaguars, thank you, Alan."[4]

Any book about jaguars must address the invaluable work that Rabinowitz did on behalf of the New World's most dominant apex predator. He considered the jaguar, the third-largest cat in the world, to be the epitome of wildness, tenacity, and resilience, capable of reshaping itself to fit a wide variety of environments, with a heroic will to live. Searching for words and failing to describe its extraordinary nature, he ultimately resorted to calling it "jaguarness."[5]

Today, almost one million years after they appeared in the New World, jaguars are spread throughout Mexico and Central and South America. Threats to their existence are growing. Yet many have guarded hope for the species.

Hope can be a powerful force. But the truth is that the jaguar is already extinct in two of its historical range countries: Uruguay and El Salvador. Today it is classified as Near Threatened on the Red List of the International Union for the Conservation of Nature (IUCN, the global authority on the status of endangered animals). However, most jaguar subpopulations outside the Amazon and the Brazilian Pantanal, the earth's largest tropical wetland, are considered endangered or critically endangered. Two main studies estimate the size of the jaguar population across its range. One approximates the population at 64,000, while the other estimates the number at 173,000. Many jaguar biologists believe that the lower number is the more realistic one.[6]

Jaguars are indeed fighting an uphill battle. In the past sixty years alone, a period some call the Great Acceleration, human impacts on the environment have unfolded exponentially, multiplying in a way that makes it impossible for most species to adapt. Currently, a quarter of all mammals, and an eighth of all birds, are classified as threatened with extinction. Montanan David Quammen maintains that by the middle of the twenty-first century, all wild predators will disappear from the Earth. Many say the Earth needs healing, but healing requires time, and as the author and environmentalist Rachel Carson put it, "In the modern world, there is no time."[7]

Alan Rabinowitz recognized, too, that time was not on our side.

He knew that if we were to save the jaguar, we would have to act fast. He argued that by protecting and restoring jaguars—a leading "indicator," "flagship," "umbrella," or "keystone" species—throughout their range, by preserving and expanding the natural corridors that reinforce vital genetic diversity, we could enhance biodiversity and perhaps also safeguard ourselves from the ravages of climate change.

Heart of the Jaguar is about Alan Rabinowitz and the apex predator that captured his imagination. To research the book, I traveled across the Jaguar Corridor, a swath of land that unites ecologically important jaguar populations and that extends roughly five thousand latitudinal miles from the Iberá wetlands in northern Argentina to southern Arizona. My journeys were fascinating, uplifting, and profoundly disturbing. I emerged disheartened by the callous destruction of our natural world, the rampage of industrial agriculture, the devastating fires fueled by climate change, the scourge of wildcat mining, the indifference of political leaders, and the growing and ghastly trade in jaguar parts. But what impressed me mightily was the dedication of so many people who carry on Alan Rabinowitz's legacy: the wildlife biologists, government officials, rangers, park and reserve administrators, conservationists, wildlife veterinarians, guides, Indigenous organizers, ranchers and farmers, journalists, community organizers, and teachers who, gifted with the fierce heart of the jaguar and a bountiful love for the species, continue to make enormous personal sacrifices to ensure that jaguars, the most magical of all the big cats, will always have a home on this planet.

Heart *of the* Jaguar

INTRODUCTION

The Rock Star

WHEN THE JAGUAR THAT CAME TO BE KNOWN AS EL JEFE crossed the U.S.-Mexico border in 2011 and entered Arizona, he had no idea he was walking onto a stage in which he would be the star performer.

He had been on the move for days, or possibly weeks, having left his home, 125 miles south of the border, in the ninety-square-mile Northern Jaguar Reserve in the Sierra Madres of northwestern Mexico, where teams of American and Mexican conservationists were struggling to protect a waning population of the world's northernmost jaguars. El Jefe had likely fled to save his life. At two years of age, and weighing just 120 pounds, staying put had become dangerous because larger, more territorial males, unwilling to tolerate an adolescent intruder, prowled the countryside. But unlike the others, who escaped with their lives by traveling south, El Jefe responded to the magnetic pull of his internal compass by fleeing north, becoming just the fourth documented jaguar in two decades to make the border crossing into Arizona and what was once jaguar country. There he was utterly alone—perhaps the loneliest jaguar in the history of the species—but content for a time.[1]

In 1963 the last known female jaguar in the United States was killed by a hunter at nine thousand feet in the Apache-Sitgreaves National Forests in the Gila watershed in eastern Arizona's Mogollon Rim.* The hunter, who thought he was zeroing in on a big bobcat, shot her at eighty yards in the evening's dwindling light. Two years later the last legally killed jaguar, a male, was shot by a hunter in the Patagonia Mountains of southern Arizona. In 1969 Arizona finally outlawed most jaguar hunting, but with no known females roaming the state, the prospects for a rebounding population of indigenous big cats were exceedingly dim.

No one knows exactly when El Jefe entered the United States, but it's possible that a Border Patrol helicopter pilot who had reported seeing a jaguar in the Santa Rita Mountains in June 2011 was the first one to spot him. The first documented sighting of El Jefe was in November 2011, east of the Santa Ritas in the Whetstone Mountains, fifty miles north of the Mexican border, by an Arizona man named Donnie Fenn. Fenn ran a part-time business, Chasin' Tail Guide Service, that specialized in mountain lion hunts. He and his ten-year-old daughter Alyson, already an accomplished horsewoman, left their home on a Saturday morning bound for the Whetstones.

Once they saddled their mules, they rode with a pack of eight dogs up ahead singing. They were about to call it a day when the strike hound cut a trail. The Fenns followed the dogs as they ran through the canyon in full cry. After a short chase, Fenn saw his hounds one hundred yards ahead, bellowing frantically. Calming his mule, he approached slowly. Assuming his dogs had treed a mountain lion, he used his telephoto lens to zoom in on the tree. As he got closer, he saw it. He was astounded by its size. He guessed that it was twice as big as a mountain lion. Then he dismounted. His heart was pounding, but he was mesmerized. He could sense the jaguar's power. He watched as the big cat slowly climbed down the tree. When its feet

---

* Babb et al., "Updates of Historic and Contemporary Records" suggest that they have "reliable information" that the female jaguar shot in 1963 had been "translocated from what is now Belize" and released on the Fort Apache Indian Reservation. It goes on to say that the last documented female was killed in Arizona in 1949.

hit the ground, he thought for a moment it might turn to pounce on him. Instead, it took off at "lightning speed" in the opposite direction, and his dogs gave chase.[2]

About two miles down the trail, the hounds brought the jaguar to bay. It was growling in a raw and aggressive way, making sounds Fenn had never heard from an animal. His hounds, excited to the point of frenzy, had encircled the jaguar, and it was swiping at them. He worried that the big cat would tear his dogs apart. But the jaguar made its final escape, and Fenn shouted to his crazed hounds, commanding them to hold hard.

"It's the most amazing thing that's ever happened to me," Fenn later told the *Arizona Daily Star*. "I got to see it in real life, my daughter got to see it, but I hope never to encounter it again."

Fenn alerted the Arizona Game and Fish Department (AZGFD) officials, who found hair samples left behind by the animal and obligingly told his tale to the media.

What El Jefe did after encountering Fenn and his hounds, no one really knows. Having learned from his experience in the Whetstones, he likely grew even more stealthy, moving largely at night. He wasn't seen again until late 2012, when some of the hundreds of remote wireless infrared cameras operated by the University of Arizona's Jaguar Survey and Monitoring Project captured photos of a male jaguar lurking in the Santa Rita Mountains. The rugged "Sky Island" ranges, like the Santa Ritas, were once the hideouts of Geronimo and Cochise and the Chiricahua Apaches who were relentlessly pursued by Generals Crook and Miles and thousands of U.S. Army soldiers. They were also the perfect hiding place for a lone jaguar.

The wild Santa Ritas, just twenty-eight miles southeast of Tucson, had everything El Jefe needed: high country, with its narrow ridges and steep slopes, swathed in pine forests, where he could move unseen; mid-elevations scattered with live oaks, Gambel oaks, and junipers; and fecund washes thick with walnuts, hackberry, willows, and cottonwoods. And there was an impressive array of wild game for the taking—deer, javelinas, coatis, skunks, turkeys, jackrabbits, and raccoons. He was able to find water in streams, ephemeral ponds, and springs. Mountain lions inhabited the area, but they seemed to know, instinctively, who the boss was.

The primary danger for El Jefe was that he was fond of space and was partial to wandering. In his next encounter with people—or perhaps a busy highway—he might not be so lucky.

Among the handful who knew of El Jefe's presence, no one was better acquainted with the jaguar than wildlife biologist Chris Bugbee. Along with his sixty-pound Belgian Malinois Mayke—who had failed as a Border Patrol drugs and explosives dog and whom Chris had carefully trained to be a jaguar scent detection dog—the burly and rugged Bugbee had been steadily tracking El Jefe's movements across the Santa Ritas for the University of Arizona's Jaguar project.

It took a long time for Bugbee to turn a skittish dog into a dependable jaguar scat detection dog. But at some point, Mayke blossomed. She'd scent over puma scat and would drive off impudent black bears, but when she found jaguar scat, she would stand over it and bark repeatedly. The longer Bugbee and Mayke tracked El Jefe, the more curious the big cat became. Often, when they would return to a camera that they had checked days before, the time code would show that El Jefe had visited that same camera just minutes after they had. In other words, the big cat was tracking them. At first, it made Bugbee break out in an anxious sweat, but as they became more familiar with each other, it turned into a "special relationship," one that Bugbee came to "cherish"—just him, his faithful dog, and the United States' only wild jaguar.

By 2015, El Jefe—though not yet enmeshed in a thicket of complicated and conflicting human ambitions—was becoming a local legend and something of a household name as media attention to the "Santa Rita jaguar" or "America's last jaguar" intensified. But the charismatic big cat was still an anonymous jaguar that had taken up residence two hundred miles north of where he was born. For some biologists, his presence was a thrilling development, signifying the auspicious return of jaguars to an area where for millennia big cats had thrived. Alan Rabinowitz, however, was not especially impressed and went on record saying that a lone big cat, especially a male, wandering the mountains of southern Arizona, was nothing more than a fortuitous exception and had little or no ecological significance. If anything, he argued, jaguars dispersing from a fragile population in Sonora, Mexico, were acting as a desperate organism might, searching for a way to survive.³

The Center for Biological Diversity wasn't buying Rabinowitz's indifference. The gutsy and contentious Tucson-based nonprofit environmental organization, with a reputation for filing lawsuits based on the Endangered Species Act since its founding in 1989, began a campaign in May 2015, focusing on the big cat and on Arizona jaguars in general. Not long afterward Mike Stark and Russ McSpadden, from the center's communication department, and Randy Serraglio, a magnetic and outspoken conservation advocate for the center, began laying plans for a considerably more aggressive publicity campaign whose ambitious goal was to brand jaguars as icons of wildness in southern Arizona. But perhaps their greatest hope, one they were reluctant even to whisper about outside the confines of the conference room, was their desire to turn the Santa Ritas jaguar into a national cynosure and, dare they hope, even a rock star.

While brainstorming, Serraglio, McFadden, and Stark hit on the notion of holding a naming contest for the big cat. An anonymous apex predator on the loose in Tucson's back forty was exciting, but a jaguar with a resonant moniker could be a powerful symbol. Further refining the idea, they decided to enlist the help of local schoolkids. They settled on the Valencia Middle School in Tucson, composed largely of Indigenous and Mexican American students, which had a jaguar as its mascot. Serraglio took the lead, contacting Valencia's principal, who embraced the idea. Together she and Serraglio established a jaguar curriculum. Following the study unit, their plan was to hold a schoolwide vote to determine the jaguar's name. Simultaneously, they ran an online vote for their members and supporters across the country, using the teaser: "Cast your vote: ProtectOurJaguars.org."

The program was even more successful than Serraglio imagined. On the final day of the study unit, the school staged a huge pep rally, replete with a large, twelve-by-fourteen-foot Chinese-dragon-style jaguar puppet, manned by five people; music and singing; and a relay race, where kids, emulating jaguars, had to secure "resources" around the school grounds while avoiding "threats" (mine, roads, and the border wall). Serraglio described the celebration as a "jaguar frenzy." McFadden filmed and also did interviews with the students about what name they chose and why. Later he spliced together the video clips, which the center used on its website.

In early October 2015 Serraglio tallied all the votes from the school and the online campaign. The top five names were: O'oshad (the Tohono O'odham word for jaguar); Rito, in honor of the Santa Ritas; Scout; Spirit; and El Jefe. The winner was El Jefe, the Boss, by a whisker. One month later, on November 2, Serraglio and the center staged a live press event at the school to announce El Jefe as the winner of the jaguar naming contest.

Meanwhile, in the months leading up to November 2, Chris Bugbee had grown frustrated with the University of Arizona's resistance to making public his stirring and unprecedented footage of El Jefe. From his perspective, the footage was "gold" for jaguar conservation in the United States that the directors of the project refused to use. He'd also become deeply upset with the Forest Service, which had issued a preliminary permit for a copper mine in the Santa Rita Mountains that would challenge the inviolability of the Endangered Species Act, which protected jaguars like El Jefe and their habitat. One of Bugbee's videos showed El Jefe just a half-mile from the proposed mine site.

Bugbee later spoke of his frustration: "We wanted to show the world that we still have jaguars in Arizona. We wanted to get the American public involved in this question: Do we want to recover jaguars, or do we want them to just become a piece of local history? Nobody wanted to do any advocacy for jaguars or say a word against this mine . . . not the university, not the wildlife agencies. El Jefe was like a dirty little secret they wanted to keep quiet. It didn't sit right with me. I felt I had to do something."

As Bugbee lamented the university's decision to keep El Jefe's existence quiet, he was almost certainly recalling the enduring taint of the Macho B scandal. In 2009 department employees had secretly snared, darted, drugged, and collared a vulnerable old jaguar. After being caught and released, Macho B fell gravely ill and ultimately had to be recaptured and euthanized. The AZGFD then orchestrated a cover-up that would eventually become public knowledge, thanks to the dogged work of two local newspaper journalists and a lone whistleblower.

So Bugbee and his wife, Aletris Neils, a fellow biologist who had studied black bears in Florida, contemplated going public with the photos and the video footage of El Jefe that Bugbee had collected under the aegis of the University of Arizona's Jaguar Survey and Monitoring Project. Though he hadn't intended to mix science with advocacy, both he and Neils sensed that the big cat could be a galvanizing publicity tool in the fight against the copper mine. They also feared for him. Rural Arizona could be hostile territory for a jaguar. Someone filled with hate and hostility might try to track down El Jefe, another unwanted migrant from south of the border, and shoot him.

Bugbee and Neils agonized over the decision, knowing, too, the kind of discord it could create in the academic and conservation community. Some would applaud it, but most would regard the move as roguish, and it would alienate colleagues who believed that the university had proprietary rights to the footage and that under no circumstances should a vulnerable jaguar's movements ever be made public. The University of Arizona and the USFWS would eventually employ a federal agent and a prosecutor to investigate Bugbee and Neils's conduct for allegedly stealing intellectual property and cameras; five years later it ended with no convictions.

At the same time, the University of Arizona's money for Bugbee's camera program ran out. Bugbee tried to convince the USFWS, the AZGFD, and even the Forest Service to pick up the funding, but when they balked, he approached the Center for Biological Diversity. The center ponied up the money to keep Bugbee's research going, but it negotiated a promise from him that he would provide footage it could use in its ongoing PR campaign for jaguars.

Bugbee and Mayke then resumed tracking El Jefe, using an

expired university research permit. By the fall of 2015, Bugbee realized that after appearing regularly and reliably on cameras for several years, El Jefe was nowhere to be found. Mayke hadn't barked in months.

In late January 2016 Serraglio alerted Bugbee that McSpadden had cut a forty-one-second video clip, which the center was planning to release on the Conservation CATalyst website, an organization Bugbee and Neils had established hoping to educate the public about big cats and promote their protection. The center had picked the day, February 2, which would precede the Barrio Brewing Company's much-anticipated rollout of its El Jefe Hefeweizen, infused with catnip. Bugbee and Neils held their breath.

On February 2, according to plan, the center released the video, actually a compilation of three separate videos, showing El Jefe, now a robust 150-pound male in his prime, moving like a ghost through the forest and up a creek bed. Midway through, El Jefe walks, broad-shouldered and muscular, right into the camera.

According to Bugbee, once the video hit and legend merged with reality, "all hell broke loose." The following weeks were nothing but a "blur." The press coverage was preponderantly positive, but the USFWS field supervisor for the Southwest Region, unaware that El Jefe had likely decamped for new territory, accused Bugbee of blatantly violating the terms of the research permit and endangering the big cat's life.

The video went viral. Twenty-one million viewers in the United States saw it, and El Jefe, the incarnation of beauty and wildness, became an object of adoration and admiration, a national—and international—sensation. He was Arizona's version of the iconic Yellowstone wolf, O-6, or Los Angeles's beloved puma, P-22, and as close to a natural cause célèbre as the state had ever seen. In the forty-eight hours after the video was released, a TV monitoring service used by the Center for Biological Diversity found it had appeared on 830 TV segments in the United States, including ABC's *Good Morning America*, NBC's *Today Show*, and *CBS This Morning*. Millions more saw the video on national newspaper websites across all seven continents.

The center's rough estimate was that 100 million people globally viewed the video. El Jefe was indeed a rock star.

But by the time El Jefe became famous, he was already gone. He likely did what jaguars sometimes do—disappear. Some claimed he went east into the Patagonia Mountains. Others said he followed his sun compass hundreds of miles back into the foothills of the Sierra Madres. Bugbee, who in almost four years of tracking El Jefe never had the good fortune of laying his eyes on the big cat, recalls some of the last photos of El Jefe, in the fall of 2015, his testicles bulging. He thinks the big lusty cat, obeying a biological imperative, went south in search of a mate, knowing that his sojourn in Arizona was reproductively doomed.

What exactly happened to El Jefe, no one knows. But for a few years he graced Arizona with his presence, and for a brief time, he captivated a nation.

CHAPTER 1

The Jaguar Craze

FROM THE PLAINS OF KANSAS TO EL JEFE'S STOMPING grounds in the rugged borderland region of the United States and northern Mexico to the Colombian Amazon and the heat-baked Gran Chaco of southern Bolivia, Indigenous, precolonial people inhabited the same landscape as jaguars. They shared their world with them, feared them and worshipped them, incorporating the jaguar's image into their architecture, art, iconography, mythology, and religion. Our relationship to the jaguar is an ancient, intimate, and mythic one, so durable, in fact, that for thousands of years the great carnivore achieved a consequential presence in the lives of many different peoples and cultures separated by thousands of miles. Metaphysically speaking, no animal was more important than the jaguar.*

Many tribes of the Amazon inhabited a world where the boundary between humans and the animals of the rainforest blurred, and they developed a complex animism to account for that; no animal was more intrinsic to their beliefs than the jaguar. In the tropical lowlands

---

* "Motifs of large felids are one of the several items separating Middle American cultures from the ursine-oriented peoples inhabiting more northern regions." Brown and López-González, *Borderland Jaguars*, 67.

of Bolivia, a Tsimane myth told of a *tigre gente* (jaguar people) who possessed the supernatural ability to shape-shift and transform into jaguars. Farther to the north the legendary Maya revered the jaguar. Because no other animal possessed its strength, stealth, and sense of caution, the Maya believed that the jaguar protected the sun on its journey from day to night and back to day. Farther north, in the desert Southwest of the United States, Pueblo lore told of the *rohona*, a "big cat with spots," that was venerated as a spirit hunter to which the Pueblo prayed at tribal ceremonies.[1]

With the arrival of Europeans, jaguars came into close proximity with people. The colonists' vulnerable livestock provided jaguars with ridiculously easy targets. Rafael Hoogesteijn, Latin America's leading jaguar conflict specialist and an expert on livestock depredation and founder of Panthera Brazil, equated a jaguar taking a calf to humans eating at Burger King. It was as easy as buying a Whopper. But as simple as it was for a big cat to stalk and kill defenseless livestock, cattle killing never became endemic among jaguars. According to A. Starker Leopold, the zoologist and conservationist and son of Aldo Leopold who studied jaguars in northern Mexico, cattle killing was rare among jaguars and was confined to areas where they lacked available prey, or to big cats whose hunting skills had been compromised by injury. U.S. president Theodore Roosevelt also weighed in on behalf of the jaguar. Roosevelt, who hunted jaguars on horseback in the Brazilian Pantanal and developed a complicated but lasting respect for the big cat, claimed that jaguars didn't bother with livestock because they preferred wild prey.[2]

In the United States, jaguars once inhabited a variety of ecosystems, from West Texas to Colorado and Missouri to Alabama. However, the first homesteaders, clinging to an irrational fear of the wild that cast predators as bloodthirsty killers, brutally disrupted an ecosystem that had thrived for ten thousand years.[3] The mind-boggling number of animals that had flourished in the wake of the Pleistocene extinctions began to disappear. Predators vanished first. Half a million gray and red wolves and a hundred thousand grizzlies. Then other animals vanished—herds of bison that had once extended to the curve of the Earth, staggering numbers of pronghorns, wild horses, elk, and bighorn sheep.[4] Soon much of the spectacular wilderness, with its "vast multitudes" of animals that John James Audubon praised so ecstatically during his trip through the West in the summer of 1843, was made safe for livestock.[5] Dan Flores writes in *American Serengeti* that the American West became a veritable "slaughterhouse."[6] Animals were massacred in what he considers the "largest wholesale destruction of animal life . . . in modern history." Although exact numbers are unattainable, the author Barry Lopez lamented the breathtaking and deliberate extirpation of 500 million animals.

Predator hunters from the U.S. Bureau of Biological Survey, whose mission changed from a purely scientific one to one of outright predator control, pursued coyotes, wolves, mountain lions, jaguars, and bears with a particular vengeance in what amounted to a war of extermination, driven by what Barry Lopez called "colonial indifference." They were numb to the consequences of their actions. In fact, the agency's merciless motto was to "bring them [predators] in no matter how."[7]

The bureau even got into the business of manufacturing poison. It built a plant in Denver, which it named the Eradication Methods Laboratory, then authorized its predatory and rodent control hunters to set out millions of toxic bait stations, including horrendously cruel M-44 "cyanide bombs" that killed tens of thousands of targeted and unintended animals and subjected them to agonizing deaths.[8] Private bounty hunters, state predator agents, stockmen, and even veterinarians contributed to the carnage. In Montana, veterinarians were required by law to infect captured wolves with mange in hopes that when they were released, the disease would spread like wildfire.

In 1917 the Arizona branch of the Bureau of Biological Survey, operating as an arm of the livestock industry, announced, with lethal intent, that "all Lobo wolves and jaguars will be taken as soon as they enter this State from Mexico and New Mexico, as one hundred percent of them live on livestock and game."[9] In 1929 the state of Arizona classified jaguars as predators, meaning that anyone could kill them at any time. Eventually, apart from a small relict population that held on in the desert Southwest and in the rugged mountain forests of central Arizona and New Mexico, jaguars were almost eliminated from the United States.

The crusade went unchallenged until a meeting of the American Society of Mammalogists, a collection of biologists and naturalists that included Aldo Leopold, Olaus Murie, and Joseph Grinnell, cast doubt on the country's policy of predator control—or more accurately, predator annihilation.[10] Studies, they said, showed that contrary to popular belief, functioning ecosystems depended on predators. If predators were fulfilling their roles, ecosystems flourished; and if they weren't, ecosystems suffered.* The Bureau of Biological Survey's E. A. Goldman, a mammalogist who led the bureau's big game and bird department, treated the message as a declaration of war, proclaiming, "Large predatory mammals, destructive to livestock and to game no longer have a place in our advancing civilization."[11]

The hate wasn't confined to the United States. Across the Jaguar Corridor, big game hunters, ranchers, farmers, and everyday people who came into contact with jaguars killed them with impunity. But it was nothing compared to the coming onslaught.

In 1962 Jacqueline Kennedy, the photogenic U.S. first lady, stepped out of a car and flashed her beguiling smile, and the world swooned. Having already charmed Charles de Gaulle, the French prime minister, and much of Europe with her beauty and grace—the French people shouted " 'Vive Jacqueline!" and the mass circulation *France-Soir* wrote ironically of "Jackie" and "the likeable young man accompanying her"—she was off on a goodwill tour of India and Pakistan.[12] For the trip, Oleg Cassini, her official couturier, who had designed clothes for Hollywood movie stars, outfitted her in a $20,000 double-breasted, knee-length Somali leopard coat with six large black buttons, a black Bergdorf Goodman pillbox hat perched on her bouffant

---

* The meeting was held in 1929. Leopold realized that the problems facing the Kaibab deer population, which was destroying the landscape, could be connected directly to the elimination of predators.

hairdo, and black elbow-length gloves. Cassini, a flamboyant and debonair ladies' man, preferred the dramatic, but Jackie often insisted on a simple but elegant style appropriate for her role as the president's wife. Nevertheless she exuded glamour, and as her fame reached dizzying heights, the "Jackie look," as it would come to be called, swept America's multibillion-dollar fashion industry and was copied by women across the globe. Few outfits, however, captivated people as much as her leopard coat, and the look spurred an international craze and a $30 million industry in exotic spotted-cat—mostly leopards, jaguars, and ocelots—coats and accessories. Cassini came to regret the fad he'd helped to start and became fervently anti-fur.[13]

Tens of thousands of hunters across Central and South America answered the call for the spotted pelts, tracking big cats unsparingly and with pathological dedication across jungles, canyons, and scrub forests. This period, built on the backs of dead jaguars, was known as the "jaguar craze," the *tigrilladas*, in Latin America.[14]

The *tigrilladas* had a disastrous effect on jaguar populations. Between 1962 and 1975, an estimated 180,000 jaguars were killed in the Brazilian Amazon alone, a number that eclipses even the most aggressive estimates of today's entire jaguar population.[15] Across jaguar territory, people chased and stalked the big cat with the same ravenousness that Clovis hunters had pursued the North American mammoth. The beautiful jaguar seemed doomed by the deadly intersection of fashion and commerce. Because jaguars, like other large-bodied carnivores, have relatively slow reproductive rates, an average of only one cub every two years (and a fairly long gestation period of one hundred days) and a protracted post-weaning period, the jaguar craze pushed the big cat to the precipice of extinction.

In 1975 the Convention on International Trade in Endangered Species of Wild Fauna and Flora (CITES) stepped in to save the jaguar, when it listed it under Appendix I, its highest level of protection, reserved for imperiled species in danger of extinction. Signed by nearly every country in the world, CITES is the primary weapon against wildlife trafficking. It was first proposed in 1963, and in the twelve long and deadly years that it took to implement the convention, hunters, eager to supply the seemingly insatiable fashion industry, continued their killing spree.

CITES is a nonbinding international agreement, lacking an enforcement mechanism, and from the very beginning, countries violated it. But whatever its drawbacks were (and are), CITES was the first meaningful attempt to protect endangered animals and plants. Before it existed, the international wildlife trade was a rapacious free-for-all of unrestricted killing and commerce. John Polisar, formerly of the Wildlife Conservation Society, has probably spent more time in the Jaguar Corridor than any biologist alive today and was a key member of Alan Rabinowitz's Jaguar Corridor Initiative team in the early 2000s. He says CITES was a desperate attempt to keep the species alive. "All those guys in Nicaragua, Honduras and elsewhere," he has said, "who had paddled their pirogues into the jungle to go kill a jaguar and sell the pelt, were slowly put out of business. CITES, eventually, shut the trade down."

CITES was imperfect in its implementation, but its impact on the international commerce in jaguar hides reduced the motivation to kill jaguars, at least temporarily. At the local level, however, it did not have the same chilling effect. The domestic trade in jaguar parts persisted.

George Schaller, the world-renowned wildlife scientist who in the late 1970s conducted research on jaguars in the Brazilian Pantanal for the New York Zoological Society (which in 1993 would change its name to the Wildlife Conservation Society, WCS), wrote that despite CITES, jaguar skins were selling for "about 3,000 cruzeiros (equal to about $100)," while ocelot skins were selling for half that, and "professional hunters guid[ed] foreign clients on illegal shoots."[16] Brazil may have been a signatory to CITES, but just across the Paraguay River in Bolivia, Schaller noted, an opportunistic entrepreneur had "opened a tanning factory to handle the many skins smuggled out of the Pantanal . . . by hunters in dugouts" who "penetrate the swamps and return laden with hides." He added ominously that the enforcement of CITES was strongly discouraged. "Frontier law," he wrote, "still operates in the sparsely settled Pantanal; piranhas are adept at disposing of dead bodies."

Belize was another place where jaguars continued to lose their lives. In the early 1980s the young biologist Alan Rabinowitz, working for the New York Zoological Society, went to central Belize at George Schaller's urging to study jaguars. There he lived something of a magical life. He walked literally in the tracks of big cats, compiling the most thorough but alarming picture of jaguar life ever produced. Despite CITES and a national wildlife protection act, jaguars were being shot, poisoned, run down by trucks, and chased with dogs by trophy hunters. Their numbers were plummeting. In an alternately arduous and uplifting two-year period, which remains a milestone in the world of jaguar research, Rabinowitz battled poachers, tropical diseases, and parasites, survived a plane crash, and performed the mundane and sometimes dangerous work of a field biologist. He diligently recorded facts and observations in his journals, all the while assembling a written history of the jaguar. "It didn't take a brain surgeon," he said, "to see the writing on the wall."[17]

CHAPTER 2

Bolivian Cocaine

ONCE IT BECAME PUBLIC KNOWLEDGE THAT EL JEFE HAD left Arizona, many biologists feared that if he had crossed the border into Mexico, he might be in imminent danger of being shot. Although the Mexican government had classified the jaguar as an endangered species in 1980, it persisted in issuing special permits to big game hunters. And while categorizing jaguars as a "priority species for conservation," it simultaneously allowed cattlemen to shoot jaguars suspected of "menac[ing] their economic possessions."[1]

In 2007, in Sonora, Mexico, Naturalia, a nonprofit conservation organization, and the Northern Jaguar Project tried to put an end to the killing of jaguars by instituting a unique program. Its aim was to enlist ranchers, for whom shooting predators was second nature, in protecting wild cats. Called Viviendo con Felinos—Living with Cats—it paid ranchers, conditioned by centuries of vigilantism, for each photo of a wild cat they produced. Despite its success, big cats were still dying across the country. But as dangerous as Mexico could be for a big cat like El Jefe, elsewhere along the Jaguar Corridor, jaguars were in far greater peril. People were shooting them not only out of hate and habit but for profit.

In 2018, in a piece for *Mongabay*, an online environmental science and conservation publication, the award-winning Bolivian environmental investigative reporter Eduardo Franco Berton set out on a daring journalistic mission across Bolivia, Brazil, and Peru to chronicle the resumption of the jaguar parts trade. He risked his life to document how the market worked and how buyers, sellers, and smugglers were operating largely unchecked and unafraid of the poorly equipped law enforcement officers who enforced the wildlife laws.

In the department of Beni in Bolivia, Berton spoke with a man who made his living as a hardened jaguar and panther hunter whom cattle ranchers called upon to do the perilous and illegal work of killing big cats. The man admitted that he was motivated by hate. As a young hunter, he was pursuing a jaguar when it turned on him. He escaped, surviving the attack, but in an act of sustained revenge, he set out to kill as many big cats as he could, and he made good money doing it. Proudly and without inhibition, he claimed to have killed thirty jaguars and a commensurate number of pumas. He told Berton that there had been a time when he would have kept the head and skin as trophies to display in his small house, as monuments to his prowess and courage, but because of the trade, he now sold everything—head, skin, claws, fangs. Buyers, especially Chinese buyers, loved jaguar parts. Some wore jaguar teeth and claws as symbols of fashion and privilege. And traditional Chinese medicine used jaguar bones to help cure an assortment of ailments. "They pay good money," the hunter told Berton. "Between 2,000 and 3,000 bolivianos [between $287 and $430] for the four fangs."

While in Beni, Berton also spoke with what was perhaps Bolivia's only female jaguar hunter. Sitting across from her, as she chewed a wad of cocoa leaves and spat liberally and without reserve, Berton could feel her energy. She was crude, tough, and fit, a cold-blooded killer, in Berton's estimation, who had to prove her ferocity and skill before local ranchers would hire her to dispose of a problem big cat. Though she wasn't as forthcoming about selling jaguar pelts and fangs, she did admit there was a local trade and that she was familiar with those who were buying and selling body parts.

In Iquitos, Peru, the largest city in the Peruvian Amazon, accessible only by air or water, Berton investigated the notorious Belén outdoor market, located on the bank of the Itaya River, a sprawling sixteen-acre maze of tables, tarps, umbrellas, chairs, tuk-tuks, and commotion, where vendors, with more than two hundred species of wild animals, living and dead, common and mysterious, legal and illegal, sell everything from stuffed boa constrictors to bottles of pink dolphin genitalia. Entering the bazaar, Berton heard the strident cries of a tiny tethered titi monkey. The air reeked of wild meat, rotting in the sun, and overhead expectant vultures clung to powerlines. He wandered around, guarded by a nervous local policeman, while vendors watched him with suspicion.[2]

As Berton delved deeper into the market, into what is known as the lower zone, the tenebrous barrio, he found noisy vendors eager to sell terrified parrots, threatened red howler and tamarin monkeys, tortoise penises, butchered armadillos, the paw of a three-toed sloth, and endless tables of smoked bush meat. In the "underground" venues, jaguar parts were brought in by motorized canoes (*peke-pekes*) from Tres Fronteras, a region 230 miles downriver where Brazil, Colombia, and Peru meet. Shocked by the brazenness of the merchants—Indigenous peoples can hunt wild game, but the sale of it is not allowed—Berton later interviewed Pedro Perez, a wildlife biologist at the Peruvian Amazon Research Institute. Perez told him that laws prohibiting the wildlife trade were failing because authorities tasked with patrolling the unauthorized markets often feared for their lives.

Exploring the craft shops outside the market, Berton discovered that jaguar fangs, claws, skins, and skulls could be readily found in many of the small parlors.* One vendor told him, using the expression *otorongo*, the Tupí-Guaraní word not for "jaguar" but for "beast,"

---

* A four-month study by the WCS and SERFOR (Peru's National Forestry and Wildlife Service), focused on Peru, revealed a more robust illegal trade in jaguar parts than previously thought. During visits to twenty-one locations in Iquitos, researchers found ninety-six jaguar parts for sale. In his article for *Biological Conservation*, Mathias Tobler writes, "While the Peruvian law prohibits any killing of jaguars and all trade with jaguar parts, there is little enforcement and teeth, claws, skin parts and even whole skins are often seen for sale in local markets." Tobler, Carrillo-Percastegui, et al., "High Jaguar Densities."

that the "'Chinese look for the *otorongo* fangs as if they were gold.'" Another Bolivian journalist, Roberto Navia, wrote that jaguar hunters had discovered that in Chinese markets jaguar fangs were as sought after as another illicit Bolivian commodity—cocaine.

In Asia, just two fangs can sell for more than a kilo of Bolivian cocaine.[3]

In another market, Berton found a spirited woman who agreed to talk with him on camera. She explained to him how buyers, especially the Chinese and hippies who come to Peru for an ayahuasca experience, concealed jaguar fangs in dried leaves and paper, which they tucked into their clothes. She spoke freely and told Berton that she was happy to sell to them. With the extra income, she could afford to support her children. She explained further that the *chinitos* (Chinese) love them, "because when you wash the fangs, they are bright white, like little pearls."

In Iquitos, in just seven days, Berton witnessed "the sale of 44 jaguar fangs, four skulls, five skins and about 70 claws," equaling roughly twenty-four jaguars. What's more, vendors bragged that their able hunters would soon be supplying them with yet more parts.[4]

The threat to the jaguar is not just a modern phenomenon. Archaeological records show that jaguar body parts were traded between Amerindian and Caribbean societies. Georges-Louis Leclerc, Comte de Buffon, a French naturalist, noted in the mid-eighteenth century, "The jaguar is not now so common in Brasil, which is its native country. . . . A price has been set on his head; numbers of this species have accordingly been destroyed; and the rest have retired from the coasts into the most desert parts of the country."[5] Over fifty years later Georges Cuvier, another Frenchman, observed, "Many parts of South America which were once grievously pestered with Jaguars, are now almost freed from them, or are only occasionally troubled with their destructive incursions." Around the same time, Alexander von Humboldt, the German geographer, explorer, and naturalist, reported, "More than four thousand jaguars are killed annually in the Spanish colonies, several of them equaling the mean size of the royal tiger of Asia. Two thousand skins of jaguars were formerly exported annually from Buenos Ayres alone." Charles Darwin also noted the absence of jaguars from the Maldonado region of what is now Uruguay. "The jaguar," he wrote, had "been banished for some years."

The persecution of the species continued for much of the nineteenth century, and then in the latter part of the twentieth, as Alan Rabinowitz wrote, the jaguar "was about to become a hugely valuable commodity."[6] In the early 1970s raw jaguar skins, fueling the multimillion-dollar trade in big cat parts, sold for almost $200 apiece. According to Rabinowitz, "Every jaguar had a dollar sign next to the bull's-eye on its back."[7] And the trade in the Americas was not confined to jaguars alone. In 1966 an adviser to the Peruvian government reported that the skins of 891 jaguars, 15,000 ocelots, and 4,000 margays were traded at the Iquitos market.[8]

CITES slowed the slaughter, and in the following decades, the jaguar population recovered. The conservation community then focused on other more pressing threats, mainly emerging habitat loss and fragmentation, brought on by the demands of unfettered logging, industrial monocrop agriculture, and retaliatory killings by cattlemen for alleged livestock depredation. Apart from a domestic business in jaguar body parts, driven by tradition, jaguars were largely overlooked by the expanding wildlife trade. But that changed around 2010, when startled biologists and conservationists sounded an urgent alarm.[9]

CHAPTER 3

The Dream

IN EARLY 1983, AFTER A HOT, BONE-JARRING TRIP OVER rutted and washed-out dirt roads, Alan Rabinowitz skidded into Belize's Cockscomb Basin on a secondhand motorcycle, a 125 cc beater he had bought for $200 in Guatemala. When he arrived, the clutch was smoking, and he knew he was lucky to have made it. Reaching deep into his pocket, he felt for the extra bills he carried, the $500 he'd need for expenses. In what was surely the most inauspicious start ever to a wildlife study, he looked around and recalled his instructions from George Schaller, the most basic of which was to embark on his research of the jaguar and determine if there were enough big cats in the basin to justify a year-long project.

Standing in the middle of an abandoned timber camp, with its derelict shacks and buildings, he wondered what to do next. How did a fledgling jaguar biologist, with little experience in the field, begin to study an elusive species in the middle of a lush tropical forest? Drenched in sweat, with mud and road dust clinging to his clothes, he contemplated finding a stream to wash in, but first he wanted to reconnoiter. Just outside the camp, he discovered a small settlement, consisting of nine thatched huts inhabited by Maya families. Over the course of his study, they would overcome their consternation and reserve and become his closest friends. Then he returned to the camp, acquired one of the better structures under "squatter's rights," lit a mosquito coil, and happily immersed himself in its "toxic fumes."[1]

Apart from what little scientists had learned from studying captive cats, the jaguar was largely a mystery to wildlife science. George Schaller was the first biologist to conduct a comprehensive study of the jaguar. In 1976 he arrived at the Acurizal Farm, at the western edge of the Brazilian Pantanal. Not long afterward the Brazilian biologist Peter Crawshaw joined him. Crawshaw would go on to become the country's preeminent wild cat biologist, inspiring and mentoring a new generation of Latin American researchers.

Before Schaller and Crawshaw, both Alexander von Humboldt and Charles Darwin had written about jaguars. In the early 1800s Humboldt traveled throughout South America and Mexico and captured his adventures in fascinating travelogues. He also observed jaguars in the wild and made a point of explaining their behavior in a style that neither romanticized nor demonized them.

When Darwin arrived in South America, he had already read Humboldt's influential *Travels and Research* and was eager to encounter the jaguar. Although he never laid his eyes on one, while hiking in jaguar country, he did run across "indubitable signs of the recent presence of the tiger."[2] Based on those sometimes-unnerving experiences and the information he collected from others, he too wrote about the jaguar and its behaviors. In fact, he incorporated the big cat into his emerging thoughts on natural selection. After having observed a jaguar fishing, a "thoroughly terrestrial quadruped" that "took freely to water," he asks, "Who will say what could thus be effected in the course of ten thousand generations?"[3] The author and conservationist Sharon Wilcox writes that Darwin, in an early essay preceding *On the Origin of Species*, used this illustrative story "as the potential starting point for the evolution of marine carnivores, including whales."[4]

The jaguar also captured the interest of John James Audubon. In his counterpart to *The Birds of America*, which he titled *The Viviparous Quadrupeds of North America*, Audubon was the first to describe and illustrate in exquisite detail the jaguar of the United States.

Three-quarters of a century later, in the first volume of *Lives of Game Animals*, the author and conservationist Ernest Thompson Seton, relying on personal observations, hunters' tales, and the reports of Humboldt and Darwin, described the jaguar at length, lamenting that it was "nearly extinct as a resident" from "California, Arizona, New Mexico, Texas, and Colorado." Around the same time, the Brazilian zoologist Rodolpho von Ihering and the Brazilian journalist and conservationist Eurico Santos penned quasi-scientific texts on the jaguar. Much of their information came from interviewing sport hunters and reflected the hunters' biases.

In 1970 the naturalist Richard Perry wrote *The World of the Jaguar*, in which he collected anecdotal information based on accounts of "hunters, explorers, and naturalists." Six years later the Brazilian hunter Tony de Almeida (Antônio Eduardo d'Andrada Almeida) published *Jaguar Hunting in the Mato Grosso and Bolivia*, in which he described the jaguar's biology and behavior, in addition to the stomach contents of the jaguars he had shot.

But the early research was nothing compared to the scholarly volumes devoted to tigers, lions, cheetahs, and leopards. Hard scientific data on the jaguar, based on systematic studies, conducted by wildlife biologists, did not exist.

In the Pantanal, Schaller and Crawshaw discovered that jaguars were regarded as cowardly cattle killers. Backed by their hatred, stockmen prosecuted a deadly campaign against the big cats. One hunter bragged to Schaller about having shot eighty-three jaguars and admitted that the number may have been considerably higher. The utter casualness with which they killed jaguars stunned Schaller. The local people told him that jaguars should be "killed like snakes."[5]

Schaller set out to acquire as much information about the jaguar as he could. He knew that the first step toward saving a species was sound science. After ranch hands at the Acurizal Farm killed two female jaguars that he and Crawshaw were studying—the world's first to be tracked via telemetry monitoring—the two biologists relocated to the enormous Estância (Ranch) Miranda, in the southern Pantanal. Dispirited by the jaguars' deaths, Schaller feared that the big cats had been reduced to mere "stragglers" over much of the Pantanal. In his journals he wrote: "The jaguar is already extinct . . . over large parts of the Pantanal, in some areas because of systematic eradication by ranchers within the past twenty-five years. No species in which a female raises an average of only one cub every two years can stand such heavy attrition. Unless local attitudes change, only a large national park can save the Pantanal jaguar. The ostensible reason for eliminating jaguars is that they kill cattle. And, indeed, they do, although the cats account for only a tiny percentage of those that die annually."[6] Schaller noted that cattle were dying mostly from negligence and poor management.

Not long after Schaller left the Pantanal, Rabinowitz arrived in Belize. At the time, the country didn't have a national park or even a wildlife sanctuary. Jaguars were considered depraved predators, and protection of the country's flora and fauna lay in the hands of a venal political appointee whose sole job it was to sell timber concessions to the highest bidder. Belize, or what was formerly known as British Honduras, had only just won its independence from Great Britain and, with only two major roads, was largely unknown to the outside world. For over a century prior to that, explorers had described it as the "white man's grave," an exotic, lethal, and disease-ridden place, where malaria, dengue and yellow fever, and leishmaniasis afflicted many. Rabinowitz's map of Belize showed nothing more than primordial "shades of green . . . of uninhabited jungle and swamp." The Cockscomb Basin, with precipitous mountains, rising on three sides, was "untouched scientific territory."[7]

Rabinowitz's first assignment lasted just ten weeks, and during that time, he joked that the only skills he acquired were how to use a machete without chopping off a limb and how to drink copious amounts of rum with the local Mopan Maya (Indios del Monte, Indians of the Forest). Despite logging strenuous miles through the nearly intractable jungle, the big cat remained as elusive as ever. In a moment of frustration, he likened studying jaguars to "unraveling the human genome."[8]

After that first stint, he returned to the United States to deliver his report, wondering if he dared to entertain the Icarian dream of being the first researcher in the world to study jaguars in a tropical forest. He knew that if he could justify the study to Schaller and Archie "Chuck" Carr III, the assistant director of the New York Zoological Society's Animal Research and Conservation Center, he would be willing to sacrifice everything.

Schaller and Carr enthusiastically authorized the study, and in his journal, Rabinowitz noted that he officially began his jaguar survey on March 28, 1983.

In the Cockscomb Basin area of Belize, rumors about the eccentric gringo circulated widely. Some speculated he was a poacher and trader who sold wild cat pelts overseas. Others said he was there to set up a lucrative marijuana operation. Still others were convinced he was smuggling out Maya artifacts. But when they learned the truth, they greeted it with incredulity. He was there to study jaguars with the hope of providing them "a safe haven from the continued onslaught of the outside world."

Early on the local Maya warned him that his life was in danger. They said that the jaguar was an evil creature. Innately treacherous, it would silently shadow a person before pouncing. Not long after he arrived, Rabinowitz came to understand their fear. He was following jaguar tracks through the forest. When he lost them in the tangled undergrowth, he turned back, retracing his steps. To his horror, along the trail he had cut with his machete, he saw fresh jaguar prints inside his own. The jaguar had doubled back and was now tracking him. He froze. The big cat, he later wrote, was so close he could smell its musky odor and "feel the heat of its body."[9]

After weeks of hacking through the lowland forest, adorned with lianas, hanging vines as thick as ropes, and spiny bromeliads, he realized that to really discover the intricacies of jaguar behavior, he would need to capture a number of big cats and attach radio collars to their necks. That would not be an easy task. Carr's father, Archie Carr, Jr., the famous naturalist, warned Rabinowitz that jaguars were "masters of eluding observation" and "walked, unseen as a ghost" (as he also wrote in *The Windward Road*).[10]

But all Rabinowitz's early efforts at designing a suitable trap failed. That's when he met Bader Hassan, a Belizean of Lebanese descent, who offered up his pack of "magnificent" hounds.[11] Hassan used his dogs for his big game business, but Rabinowitz depicted him as anything but a heartless trophy hunter. He respected him because Hassan loved the adventure of running jaguars; he enjoyed the "chase, not the kill."

After an exhausting first day, trying to run through the nearly impassable rainforest, Rabinowitz wasn't sure he'd be able to continue. His job was to run with the sweat-soaked dog handlers, across the "rugged terrain," inhabited by venomous snakes, chiggers, flies, ticks, mites, swarming sweat bees, and mosquitoes, while carrying the drugging and radio tracking equipment. Without a trace of self-pity, he wrote that crossing the jungle was so taxing it would have been ideal training for U.S. Olympians. But the heavily muscled Rabinowitz, who stood five foot nine and a half and weighed 185 pounds and looked more like a collegiate wrestler than a steeplechase runner, continued. Somehow, despite his inner thighs being rubbed raw and his aching knees, he found the energy to run day after day, following the crazed dogs.

At night, fighting off sleep, he wrote in his field journals, developing a ritual he would follow throughout his professional life. His habit was to transcribe the notes he took during the day and then to expand on those notes, adding details and descriptions, struggling always to express his innermost feelings and emotions.[12] Schaller had stressed to him the importance of cultivating the discipline of writing, of keeping the "sights, sounds, and smells of the expedition alive" and of "leavening science with thoughts about the survival of the species." The pen, he insisted, was a "weapon against oblivion," and writing was essential to "helping species to survive."[13] Schaller felt, too, that a scientist had an obligation to communicate with ordinary people. Scientific articles, however well written, did not have the ability to engage the public.

On June 21, almost three months after arriving in Belize, Rabinowitz wrote that he woke "feeling strong and ready," which was a good thing because by six a.m., he and the agitated dogs were on the run, "up and down hills, through thickets and over large fallen trees."[14] Thorns and branches tore at his shirt and pants, and soon he was bleeding from his arms and legs, his heart nearly exploding from his chest. But he felt alive as never before. Later he would write, "It was like a drug, carrying us to a different level of consciousness, one of the most elemental levels of human existence, the hunt and the chase."

Two and a half hours after the chase began, Rabinowitz heard Hassan yell from a ridge. Separated by harsh terrain, Rabinowitz struggled to locate Hassan, and when he did, he found him kneeling next to Guermo, one of the dog handlers. A large, highly venomous fer-de-lance snake (from the French, meaning "iron spear point"), with its head nearly as big as a fist, lay dead on the ground. Guermo's head rolled back and forth, as if he were possessed by demons. Still gasping for air, Rabinowitz knelt next to him. When he lay two fingers on the inside of Guermo's wrist, Guermo recoiled in pain. Rabinowitz could feel that his pulse was rapid and faint, and his leg, where the snake had sunk its teeth into his flesh, was already beginning to swell. That was when Rabinowitz realized that he had left the antivenin back in the truck. He knew that without it, Guermo would die. While Rabinowitz tried to keep Guermo comfortable, Hassan sent one of his dog handlers back to the truck.

When the dog handler returned, Guermo was writhing and bleeding from his tongue. Soon, Rabinowitz knew, blood would seep from nearly every pore. He injected Guermo with 40 cc of antivenin, while Hassan's trackers cut poles and fashioned a crude stretcher.

By the time the men lugged him back to the truck, Guermo was delirious. Laying him in the bed of the truck, they hurried back to camp. But the road was potholed, and with every bounce and dip, Guermo cried out in pain. When they finally reached camp, Hassan had already radioed for a plane, which was waiting on the rough airstrip ready to take Guermo to Belize City.

Arriving in Belize City, Rabinowitz and Hassan delivered Guermo to the hospital. They waited and discussed his treatment with the doctors, who wanted to amputate Guermo's leg. Rabinowitz angrily insisted he be allowed to consult with physicians at the segregated British Hospital, who told him that amputation was unnecessary if antivenin was administered. Rabinowitz bullied the Belizean doctors, and when he was assured that they would follow the antivenin protocol, he returned to Cockscomb, satisfied that he had done everything in his power to save Guermo.

The following morning, back in Cockscomb, Rabinowitz checked on his jaguar bait, a pig, as he had done every morning for the past weeks. This time, however, he saw an empty rope and the pig's intestines in a pile nearby.* The jaguar was close. Although exhausted by the previous day's ordeal, Rabinowitz and Hassan knew they couldn't pass up a golden opportunity to finally tree a jaguar. They released the hounds, and the chase was on.

---

* Rabinowitz "didn't like to use live animals as bait, but since it was necessary, [he] used animals that would have been killed and eaten anyway." Rabinowitz, *Chasing the Dragon's Tail*, 41, 126.

CHAPTER 4

The God of Death

RABINOWITZ FELT THE ADRENALINE SURGING THROUGH his body, the atavism of the hunt, and the bloodlust of the dogs. After bounding up and down a series of sharp hills, he listened and knew the dogs had treed a big cat. When he walked up to Hassan, still panting from the exertion of the chase, Hassan pointed to the jaguar resting on the arm of an ironwood tree. In the dappled light of dawn, Rabinowitz couldn't see it until he fixed the beam of his flashlight on a branch.

In *Jaguar*, Rabinowitz wrote rapturously of the experience: "I was looking into the face of the animal I had sought for so long. I heard nothing but my heartbeat. I felt naked and alone, as I confronted wild, untouchable beauty."[1]

Intending to sedate the jaguar, Rabinowitz and Hassan approached the tree, careful not to spook the big cat. When Rabinowitz raised his blowgun, he was shaking with emotion. A branch deflected his dart, but Hassan's had found its mark, and the two backed away from the tree, waiting for the sedative to take effect. Five minutes later, when they approached the jaguar, it was alert and growling. Again, they stepped back, and Rabinowitz loaded another dart. This time the jaguar started climbing down from its perch. When Rabinowitz lifted his blowgun again, the jaguar suddenly leaped from the tree, soaring just above his head. Dazed, he lay on the ground, with his bent blowgun beside him, and realized the jaguar had knocked it from his hands.

When Hassan again released the dogs, they cornered the big cat in the hollow trunk of a fallen tree. While the dogs held the jaguar at bay, Rabinowitz ran to the truck to get another syringe and a rifle to replace his broken blowgun. Back at the tree, Hassan shined a flashlight into the cavity and Rabinowitz knelt at the opening just three feet from the jaguar's "angry shining eyes."[2] He aimed at the big cat's neck and fired. When he loaded the rifle with a second syringe and fired again, the jaguar rushed the opening. Rabinowitz let out an involuntary scream and rolled onto his back, defenseless. The jaguar might have pounced on him, if not for Hassan and his dog handler, who blocked the log's opening with stout wooden poles.

Next came the job of pulling the tranquilized jaguar out of the trunk. Porfilio, Rabinowitz's trusted Maya assistant—the two were nearly inseparable—got the assignment because of his size. Porfilio was barely over five feet tall. He slithered into the narrow hole, until Rabinowitz could see nothing but his shoes. Then he yelled for them to pull, and when they did, Porfilio came out with the jaguar in tow.

Rabinowitz injected the jaguar with Valium and then, knowing he had only a limited amount of time, took body measurements while Hassan recorded the details. A male, it weighed only 123 pounds.

When Rabinowitz finished the examination, he took a moment to dwell on the image of the jaguar, the first ever to be tracked by a radio collar in the rainforest, before the tranquilizer wore off. An ugly scar ran across its nose, and its ears were tattered and torn. What kind of life, he wondered, had it lived in the jungle? Up until that point, the big cat led a mysterious existence. But once it wore the radio collar, Rabinowitz knew their lives would be intertwined. He would know by the sound of the signal if it was resting or running or feeding. He would be able to enter the jaguar's secret realm.

He worried that if by fitting it with a small radio collar, weighing just half a pound, he was "affecting its chances for survival in a world already rugged and harsh."[3] He knew the collar could deliver essential information. But he couldn't help but feel he was "breaking a wild stallion." He was a scientist and was "there to try and help save it [the jaguar] from extinction . . . trying to prevent more damage by a society that was turning its back on the natural, free-living creatures of our world." He struggled to understand his turbulent emotions, the sense of triumph together with a stabbing sense of loss. He had crossed the threshold between the feral world of the jaguar and the far tamer realm of humanity, and somehow he intuited "it would never be the same." The elation he felt at being able to capture and touch the wild cat was complicated by an ineffable sense of sadness. He wrote that his "interference was soiling something beautiful."[4]

As the tranquilizer wore off, and the jaguar lurched into the jungle, Rabinowitz's thoughts turned to Guermo. If Guermo died from the snakebite, how could he ever justify his efforts?

Two days later Rabinowitz called Hassan to inquire about Guermo's condition. "He's dead, Alan," Hassan told him.[5]

Guermo had died on the very day they captured the jaguar. Guermo's wife had taken him out of the hospital and delivered him to a local snake doctor, who insisted that he never should have been taken to the hospital in the first place. Rabinowitz wrote that Hassan's men blamed him and Hassan for Guermo's death, and when Rabinowitz returned to Cockscomb and broke the news to the people, they too held him accountable. Even Porfilio grew distant and mistrustful. Days later he left the camp to return to his town near the Guatemala border. Rabinowitz hoped that he could convince Porfilio to stay, but he knew that a Maya man's first responsibility was to his family, not to the jaguar project of an entitled white man. He feared, too, that his Western arrogance had played a role in driving Porfilio away.

Guermo's death and Porfilio's departure left him feeling lonely and uncertain about himself and the project. Darkly, he named the jaguar, the animal that would help him break startling new ground in the realm of wildlife biology, Ah Puch (pronounced *ay pook*), after the Maya god of death.

ONCE RABINOWITZ'S GUILT SUBSIDED, he thrived in Belize, and Howard Quigley wasn't surprised. "They didn't call him Indiana Jones for nothing. Alan welcomed the opportunity to push himself to the limit."

But even Rabinowitz occasionally wondered what it was about Cockscomb that drew him so powerfully: his commitment to gathering essential data that might help to protect jaguars in a modern world inimical to wild creatures; the presence of the Maya families he had come to care about so deeply; and his sense of gratitude and obligation to George Schaller, Chuck Carr, and the New York Zoological Society, which was paying for the study. But it was also the sheer excitement of living dangerously. Schaller had told him that "searing heat" and "tropical diseases" were "occupational hazard[s]," and from the very beginning, Rabinowitz largely welcomed the project's physical and mental challenges, as well as the isolation.[6] Later, in *Jaguar*, he wrote that he had never "felt so alive as when [he] had to fight to stay alive." Cockscomb had "become [his] world."[7]

Another emotion was almost certainly at work. Rabinowitz, who had been treated for much of his early life as an outcast because of a speech impediment—which was still quite pronounced when he lived in Belize—never felt judged by the Maya of the camp. They didn't understand his passion for the jaguar, nor his sometimes-mercurial temper, but they embraced him without precondition.

"I was at home in Cockscomb," he wrote. "I liked living with the Indians. They accepted me without always understanding or even agreeing with my goals. That was more than most people had done in my life."[8]

As a boy with an incapacitating stutter, he shrank from human interaction. He underwent hypnosis, suffered useless shock and drug therapy, and was counseled by well-meaning speech therapists and psychologists. Still, he struggled to utter even a word. "My earliest memories are filled with pain, embarrassment," he wrote in *Jaguar*. "I was one of life's broken creatures." Friendless, Rabinowitz gravitated toward animals. His companions were "a little menagerie of chameleons, green turtles, garter snakes, and hamsters." Like him, they "had no voice."[9]

To make matters worse, home in Far Rockaway, at the eastern edge of the New York City borough of Queens, ruled by his fearsome father, whom Rabinowitz hated, loved, and idolized, could be an unhappy place, "filled with tension," where the "sound of laughter" was rarely heard. Rabinowitz's father, Frank, whom friends called "Red," was a force of nature. A former army paratrooper with the 82nd Airborne turned high school coach, he was a strict disciplinarian, cursed with his own memories and demons. He held down three jobs and loomed large in the life of all his children (Rabinowitz had two sisters), but never larger than in the life of his son. Red didn't know what to make of his damaged son, so he taught him the only lessons he knew—lessons in manliness, how to fight and box—emphasizing that life was a battle for which he had to be prepared, both physically and mentally. He built a gym in the basement of their home, and in that gym Alan learned the discipline that would bring about the transformation of his physique. Because of his stutter, he was endlessly teased and was often forced to defend himself. He never picked a fight, but as he wrote with pride in *An Indomitable Beast*, he "never backed down from confrontation" either.[10] Even when his tormentors were clearly bigger than he, and the odds were stacked against him, he relished the "physical power" of trying to overcome them.

The Lion House at the Bronx Zoo was the only place where he felt at peace. As an act of kindness, his father would take his son there once a week and allow him to roam to his heart's desire. Rabinowitz felt an immediate kinship with the big cats, especially the jaguars, because, as he would later write, he understood their pain and plight. He too felt caged, isolated, silenced.

It wasn't until he attended Maryland's McDaniel College that he developed an interest in the outdoors, using his weekends to camp and hike in the nearby hills. In his junior year, he discovered animal ecology. After only a few weeks, he recognized it as his calling. That summer, encouraged by his parents, he attended an experimental program in upstate New York for people with severe stutters, and for the first time in his life he was able to speak. He would always be a stutterer, but the techniques he learned at the clinic enabled him, finally, to articulate his thoughts. He wrote later that it was then that he committed himself to the big cats. He would devote his life to speaking for them. He would be their voice.[11]

Being able to speak eventually freed Rabinowitz from his self-doubt. After threatening, out of spite and rebellion, to leave school and hitchhike across the country, he graduated from college with a double major in biology and chemistry and did so with the highest honors. Now, confident in his abilities and possessing boundless ambition, he entered a Ph.D. program in ecology and wildlife biology at the University of Tennessee. There he met his lifelong and perhaps dearest friend and fellow wildlife biologist, Howard Quigley, as well as George Schaller, who was visiting the university. Schaller in particular would change the course of Rabinowitz's life, as he would Quigley's. While he was in Knoxville, Schaller and Rabinowitz took a providential day hike in the Smoky Mountains.

Lean and lithe, Schaller set a fast pace, all the while expounding on the similarities and differences between the Smokies and China, where he had studied giant pandas. In *A Naturalist and Other Beasts*, he wrote of his obsession with fieldwork, of the "long days and nights without sleep, tracking animals through dank or thorny vegetation," and admitted that he was "someone who suffers from culture shock not when settling into a project but on returning home."[12]

As Schaller spoke, Rabinowitz longed for the kind of life he described. When Schaller left, Rabinowitz had no inkling that their paths would ever cross again. But a week later Schaller called one of Rabinowitz's professors, and at a party that evening, the professor shared with Rabinowitz the good news: Schaller had been impressed with him and wanted to know if he would be interested in studying jaguars in Belize.

Though Rabinowitz, as he later admitted, was largely unaware of Schaller's mighty reputation, he would soon model himself after the indefatigable biologist. But Schaller was more than a scientist. He was also a published author, writing over half a dozen books about his fieldwork in far-flung places across the world, including his 1972 *The Serengeti Lion*, which chronicled the impact of the lion and other predators on the Serengeti's vast ungulate herds. That book won the National Book Award.

CHAPTER 5

The Survivor

NEARLY ONE MILLION YEARS AGO, *PANTHERA ONCA AUGUSTA*, commonly known as the Pleistocene jaguar or the giant jaguar—an extinct subspecies of the modern big cat—made its way into the New World. It ranged across Siberia's braided rivers and over the Bering Land Bridge that appeared and disappeared, Atlantis-like, with the fluctuations in ice age glacial cycles, leaving behind a Eurasian population of jaguars.[1] According to Charles Repenning, the American zoologist and paleontologist, there were three, perhaps four, major dispersals across the Bering Land Bridge. During the first one, approximately 20 million years ago, a warm and humid climate prevailed, and roving mammals would have traversed vast and verdant forests. By the third or fourth dispersal, as fossils indicate, the environment had changed dramatically. The forests had disappeared, brutal arctic conditions persisted, and a cold and constant wind that wrenched moisture from the air raked the land.[2] Mammals on the move would have traveled in both directions, but the great majority advanced west to east.[3]

The massive land bridge that extended for a thousand miles, from the Mackenzie River in Canada's Northwest Territories to the Anadyr and Kolyma rivers in Siberia, abounded with life. Its sheer biomass was nothing short of extraordinary. An astounding diversity of mammals dominated by large grazers and apex predators wandered the landscape. Jaguars found themselves competing for prey among a host of ferocious carnivores that included saber- and scimitar-toothed cats with knifelike fangs, seven-hundred-pound American lions, huge steppe lions, and false cheetahs, as well as dire wolves, enormous brown bears, and packs of long-legged hyenas.

When they reached the North American continent, the migrating jaguars soon encountered two monumental shields of ice, so immense that they engulfed the highest peaks. Stranded, they made their way back to the Old Crow Basin, an unglaciated refuge in what is today the Yukon Territories, where countless generations of their offspring were forced to wait for a warming period (the Pleistocene was marked by twenty such periods) before roaming south across Canada and into the Great Plains of the United States.* There they would discover an expansive terrain of flowing grasslands that resembled the modern savannas of southern Africa. Disbursing across much of the United States, they pressed farther south into Mexico and Central America, then across the Panamanian land bridge that had appeared millions of years before Beringia. Eventually the species traveled as far south as northern Argentina's expansive Iberá wetlands, where today a rewilding experiment is releasing captive-born-and-bred jaguar cubs into the surrounding grasslands and swamps.

As the jaguar was populating North and South America, 170,000 years ago, *Homo sapiens* was exploring its own terra incognita, spreading out across Africa for Asia and Europe. But *Homo sapiens* wouldn't

---

* The earliest confirmed fossil record of *Panthera onca* in the American continent is 850,000 to 820,000 years BP. Found at Hamilton Cave in West Virginia, it supports a hypothesis that jaguars became extinct in North America toward the end of the Pleistocene, and if not for being recolonized by jaguars that had already made their way far south into the New World, they would have remained extinct. "While fossil records place the jaguar in the Carolinas and Florida 7,000 to 8,000 years ago, significantly more recent accounts appear to locate them in eastern regions of North America as late as the eighteenth century. A map produced by Sebastian Cabot of this 'Terra Incognita' in 1544 includes a drawing of a spotted cat on the eastern coast of North America." Wilcox, "Encountering El Tigre," 69.

enter the New World for over another 150,000 years. Though debate in the archaeological community about just when people first set foot in North America persists, one certainty is that they did not bridge the gap between Asia and the New World until the final act of the Wisconsin Ice Age that closed the door on the Pleistocene Epoch. When they did make the jump, there was no stopping them.

Nearly all dates in Paleoindian archaeology are contested. The genomes of living Native Americans suggest that their ancestors first arrived in North America around 20,000 years ago (some say as many as 40,000). When the first explorers set foot here, what they discovered must have sent their spirits soaring. North America teemed, spectacularly, with large, protein-rich prey. It was far richer in fauna than it is today.

As these voyagers migrated, they brought their bone and stone tools, their fire-making and sewing abilities, and their fishing and hunting skills, which included the ability to tip their spears with knife-like points of obsidian and chert, sharp enough to penetrate thick megafauna hide. They also brought a driving curiosity and spacious imaginations, and they migrated with stunning speed, spreading east to the Atlantic seaboard, west to California, and south through Central and South America.[4]

But when the winds blew hard across the landscape, and alpha predators caught their tempting scent and tracked them, our far-ranging ancestors must have felt an overwhelming sense of terror. Danger and death lurked everywhere. In *Monster of God*, David Quammen points out that the term *alpha predator* has no basis in science. Its realm is that of the human imagination and its ability to summon dread. "What sets them [alpha predators] apart from all other creatures," Quammen writes, "and places them in commonality with one another, is that each of these species has members big enough, fierce enough, voracious and indiscriminate enough to occasionally kill and eat a human."[5] Many early explorers died terrible deaths in the jaws and claws of ferocious beasts. But their adventurousness paid off, for in a very short time, they found themselves perched at the top of the food chain.

The first migrants to North America trickled in at different times, from different areas, and used a variety of routes to explore the continent. Some, driven by wanderlust, chose the interior routes and followed pristine river systems, where they encountered colossal fish runs; others opted for the coast, an alternate migration path, which by seventeen thousand years ago was beginning to emerge from enveloping glaciers.

These pioneers fanned out across the continent, reproducing swiftly. As they spread out, they adjusted to a variety of climates and ecosystems with unsurpassed adaptability and, very quickly, unprecedented lethality. The largest of the animals—mammoths as big as gravel trucks, caribou, sheep, reindeer, pronghorn, musk ox, elk, moose, long-horned bison with a main beam spread of six feet, beaver as big as buffaloes—didn't know what hit them. Armed with spears with piercing blades, and sometimes dogs, prehistoric humans were more than just another carnivore competing for prey. They killed with a rapacity and skill that the continent had never seen before.

One family of animals after another went straight to the grave. Although they quickly learned to fear—perhaps over the course of just one generation—it was not quick enough. Hit by the dual threat of human hunting and radical climate change, the ungulates, unable to reproduce fast enough to outpace accelerated mortality, and with "no evolved defense systems," disappeared en masse. They were followed by the mass extinction of the beasts that preyed on them.

When human beings invaded the continent, they delivered the decisive blow to its Pleistocene-era beasts that otherwise might have been able to weather the radical climatic changes. They altered, then broke the food chains that had sustained the ecosystem for hundreds of thousands of years. By ten thousand years ago, not a single mammoth was to be found in the New World. In *Atlas of a Lost World*, Craig Childs asks the essential question of the early American pioneers: "If they saw the end of mammoths coming, would they have stopped killing them?"[6]

Some scientists call the knockout punch delivered by primitive peoples the "Blitzkrieg" or "overkill" hypothesis. The paleoecologist

Paul Martin, of the University of Arizona, first proposed the idea in October 1966 in the journal *Nature*: "When the chronology of extinction is critically set against the chronology of human migrations and cultural development, man's arrival emerges as the only reasonable answer."[7] Inspired in part by advances in radiocarbon dating, Martin discovered what he interpreted as a deadly intersection between the arrival of the first humans in North America, and their advanced hunting techniques and technology, and the disappearance of the great mammals, which they pushed to the brink of extinction and then over the precipice. The effects were devastating. In the two thousand years that it took for people to spread across the New World, from Alaska to Patagonia, 75 percent of the large animal species were eliminated.*

Martin's colleagues rejected the theory as preposterous, and their outrage was immediate. Even Louis Leakey, whom Martin had visited in Tanzania, took umbrage. *Nature* published their retorts. Undaunted, Martin noted that one hundred years earlier Alfred Russel Wallace had arrived at the same conclusion.

Despite the disagreement, one thing is clear: When the megagrazers tumbled over the cliff of extinction, they took with them apex predators like the American lion and saber-toothed cats. The once-invincible big cats vanished into myth.

At a time when extinction was a near certainty, the jaguar—which had prevailed through sheer tenacity or, as Alan Rabinowitz wrote, through an "audacity to survive"—emerged as one of the new era's premier apex predators, "a solitary, opportunistic, stalk-and-ambush" killer capable of climbing trees (jaguars are "scansorial") and swimming swollen rivers.[8]

Remarkably, the jaguar had managed to re-create itself. As the climate turned warmer and wetter, and forests appeared in the absence of the monster vegetarians, it became smaller and stockier. Rabino-

---

* Arthur Jelinek, a paleontologist, refers to early man in North America as a predator "against whom no [naturally] evolved defense systems were available." Jelinek, "Man's Role," 199.

witz attributed its reshaping to what is called Bergmann's Rule, a theory based on surface-area-to-volume ratio, in which animals in warmer climates are smaller because smaller bodies are able to expel heat more quickly. But the jaguar was not just a jungle creature. It had honed its survival skills and adjusted to incredibly diverse ecosystems, from humid rainforests to Sinaoloan thornscrub. It had acclimated so well, in fact, that in 1913 some U.S. states were offering bounties of up to five dollars per animal. Before it became the object of a relentless crusade of elimination, much like those conducted against wolves, grizzly bears, and coyotes, and its range contracted, it was the most "cosmopolitan" of species.*

Over 150 years ago Charles Darwin explained in *Origin of Species* the imperative that allowed the jaguar—like humans—to emerge victorious from the Pleistocene, while so many of the other, presumably fit, species perished. Darwin's idea was that a species's fitness and adaptability were the keys to its success. And no creature was as fit or adaptable as the jaguar.

---

* *Cosmopolitan* is a term used to describe species that are flexible and adaptable with regard to habitat.

CHAPTER 6

The Hunter

THE *PANTHERA*, OR BIG CAT, CLAN IS A GENUS WITHIN THE Felidae, or cat family, which consists of forty-one members. It is divided into five distinct though anatomically similar species—the jaguar, tiger, lion, leopard, and snow leopard—derived from one common ancestor.* The jaguar, which closely resembles the leopard, is the only member of the *Panthera* genus that inhabits the Americas.

Divisions within the pantherine genus follow those established by the Swedish botanist Carl von Linné, who assigned all living creatures a two-name designation based on anatomical traits. The first word in binomial nomenclature is the family name, and the second is the species name. For instance, the jaguar is known as *Panthera onca*, while the tiger is *Panthera tigris*, and the Amur, or Siberian, tiger is *Panthera tigris altaica*.

The jaguar is the largest cat in the Western Hemisphere. A highly specialized obligate carnivore (it eats only meat), it is blessed with a short face and a powerful jaw that allows its structurally reinforced and pressure-sensitive upper and lower canines to compress with

---

* The subfamily Pantherinae includes the *Panthera* species and *Neofelis*. The *Neofelis* genus includes the clouded leopard.

enormous force (nearly 682 newtons). While its forelegs and dewclaws anchor its prey, it kills by crushing the cranium; in this respect it differs from lions and tigers, which slowly dispatch their prey by strangulation. Exerting nearly fifteen hundred pounds of compressive force per square inch, a jaguar's bite is greater than that of the tiger.* Its bite force quotient is 7 percent greater than a tiger's, 18 percent greater than a lion's, and 31 percent larger than a leopard's.[1] In fact, it outweighs almost all living animals except the hippopotamus, the American alligator, and various crocodiles.[2] In *A Naturalist and Other Beasts*, George Schaller writes of the jaguar's power: "The jaguar takes the head into its mouth and, with its opposing canines, punctures the bone to the brain. The technique is noteworthy not only for the precision with which the canines pierce the skull on or near the ears but also for the strength needed to penetrate half an inch of bone."†[3]

When a jaguar makes a kill, its carnassial teeth, which are shaped like knife blades, slice through bulky muscles and sturdy tendons. A binge eater, it has a digestive system capable of processing dozens of pounds of meat. And a jaguar is not fussy about what it consumes. Historically, it survived when other Pleistocene-era felids dependent on large herbivores could not. Rabinowitz wrote that eighty-five prey species, from sloths to squirrels to frogs, have been noted in the jaguar's diet.‡ A. Starker Leopold interviewed a Mexican jaguar hunter, with sixty jaguar kills to his credit, who told him that the contents of various jaguar's stomachs had never been the same.

A superb and nimble hunter, the jaguar possesses an efficiency of design and an alertness that is almost unrivaled among the world's Carnivora (meat-eating mammals). With its tanklike construction and lack of speed and endurance, it can't chase down a deer or a javelina,

---

* "A large male jaguar is believed to exert up to nine hundred pounds per square inch of tooth." Mahler, *Jaguar's Shadow*, 12.

† Big cats strike "in an instant": as the canines "clamp the skull, the large front paws grab and twist the neck sideways at an impossible angle," and the prey is "dead before it hits the ground." Rabinowitz, *Jaguar*, 196.

‡ Kevin Seymour, a vertebrate paleontologist, also wrote of the jaguar's "85 prey species." Seymour, "Panthera onca."

but everything else about it is built for the hunt and the ultimate kill. It is able to move as much by feel as by sight. Its coat, broken up by a pattern of inky-black rosettes, enables it to slink unnoticed until it is ready to pounce. Its massive chest, limbs, and powerful hindquarters allow it to launch its six-foot frame fifteen feet into the air. Then its deadly jaws do the rest of the work.[4]

Though the jaguar isn't endowed with the olfactory capability of wolves or bears, its binocular vision and depth perception, allowing it to calculate the distance between itself and its next meal, more than compensates for its inferior sense of smell. At night, its eyes work like a camera's aperture, adjusting their diameter at low levels of light, with the pupils expanding and the light receptor rods on the surface of the retina activating to allow for image perception. It is this reflecting layer, with its *tapetum lucidum* (bright carpet) cells, that give the jaguar its signature green, glowing eyes.

The jaguar also possesses the auditory ability of the fish owl. It can detect slight movements of prey at incredible distances and can hear across an astounding array of frequencies, including those above twenty kilohertz. In other words, its *Umwelt* is nearly unbounded.

The word *Umwelt*, meaning "environment," was popularized by the German zoologist Jakob von Uexküll.[5] Uexküll believed that an animal's *Umwelt* represented its perceptual world, what it was capable of sensing and experiencing. The jaguar's gift is that it has an especially wide-ranging one. In addition to its sensory superpowers, it has the intensity, cleverness—George Schaller wrote of the jaguar's ability to "outwit" him—and curiosity to explore its environment in a way that many other animals do not.[6]

Most scientists consider opportunistic eaters, with diverse diets, to be especially intelligent. The diet of a giant anteater, for instance, is exclusive and largely stationary compared to that of a jaguar, which eats widely and must insert itself into the *Umwelt* of dozens and dozens of different prey species. Consequently, the jaguar must process an extraordinary diversity of stimuli, signals, and clues. A lone, largely nocturnal hunter, solely reliant on its own skills and ingenuity, it must adapt, think, and even anticipate.

In her 2003 article for *The New York Times*, titled "At Last, Ready for Its Close-Up," Natalie Angier wrote glowingly of the jaguar's hunting prowess: "The jaguar has evolved a two-pronged approach to fetching dinner—stay virtually invisible to the last possible moment and then deliver an overwhelming blow."[7]

It was this ability and an ingenious opportunism that allowed the jaguar to survive and flourish in the New World.

CHAPTER 7

People of the Jaguar

FOR THOUSANDS OF YEARS, PEOPLE LIVING IN THE AMER-
icas inhabited the same landscape as jaguars, incorporating the species into nearly every aspect of their lives. The Arawak-speaking peoples of the northern Amazon and Suriname believed that the world in which they lived was dominated by the spirit of Aruwa, the jaguar.

Among the Mojeños of Bolivia, shamans used their powers to communicate with the jaguar spirit. For the Desana of the Colombian Amazon, the word for shaman, *ye'e*, is the same as the word for jaguar. In northern Colombia, the Kogi still refer to themselves as the "sons of the jaguar."[1] The anthropologist Gerardo Reichel Dolmatoff writes of a Kogi myth of shamans who could turn themselves into jaguars and achieve a level of existence between the human and animal worlds made up entirely of jaguar people.

In the Guahibo tribe of the vast Venezuelan Llanos flatlands, shamans paint their faces with black jaguar-like spots, wear necklaces of jaguar teeth and ceremonial headdresses adorned with jaguar claws, and carry bags of jaguar fur and narcotics, which they use to achieve a trance state. The Sikuani, a subgroup of the Guahibo, use the hallucinogenic *yopo*, or *Anadenanthera peregrina*, to reach a half-conscious state, in which men dance and sing the words "We are jaguars, we are dancing like jaguars; our arrows are like the jaguar's fangs; we are fierce like jaguars."[2]

One of the first societies to venerate the jaguar was the Olmecs. Beginning around 1250 B.C., the Olmecs built an advanced culture, ruled by dictatorial city-state monarchies. According to the archaeologist Alfonso Caso, the Olmecs, whose civilization was a mystery until the mid-1900s, were the "Mother Culture" of all subsequent Mexican and Mesoamerican civilizations.[3] The Maya in particular emulated the Olmecs. In *Maya Cosmos*, its authors write, "Just as our culture sees the Greeks as the source of much of our fundamental culture and philosophy, so the classic Maya saw their source as the Olmec."[4]

The Olmecs called home the lowland rainforests of the Gulf of Mexico (in what is today southern Veracruz, Tabasco, and Campeche) and adopted the jaguar as a major part of their cosmology. The author and archaeologist Nicholas Saunders writes that with the Olmecs, the jaguar achieved its cultural zenith. "We are emphatically not witnessing a simple worship of jaguars," he wrote.[5] In *Borderland Jaguars*, David Brown and Carlos López-González also write of the presence of the jaguar in Olmec cosmology, suggesting that together with its "serpent counterpart, the jaguar deity formed a supernatural werejaguar that controlled life itself."[6] In *World of the Jaguar*, the author and naturalist Richard Perry describes the Olmecs as "jaguar psychotics," whose idolatry of the big cat caused some to submit to what can only be called self-torture; they allowed their heads to be reconfigured to resemble the skull of the jaguar.[7]

La Venta, an elaborate ceremonial center and site of Mexico's first pyramid, was arguably the most impressive manifestation of the Olmecs' most deeply held beliefs. There they designed three fifteen-by-twenty-foot mosaics to resemble masks of menacing jaguars. The ceremonial center contained twenty-ton basalt statues, carved to resemble jaguar heads. As he roamed its ruins, Rabinowitz marveled at La Venta. The entire site, he said, "was built in tribute to the jaguar" and celebrated "the earliest tangible relationship between man and jaguar."[8] For this reason, he called Mexico the "birthplace of the Jaguar Cultural Corridor."

Elsewhere, throughout their empire, the Olmecs built idiosyncratic stone monuments, some of which are thought to depict jaguar-people with Mongoloid-like features, resulting from sexual unions between a jaguar or were-jaguar and a woman. According to the archaeologist Michael Coe, author of *Mexico*, the Olmecs believed that the unnatural union created a superior "race of were-jaguars" that over time manifested distinctly human features.[9] The descendants of these couplings established the line of powerful Olmec royal families whose relationship to the jaguar formed the basis of their authority and power.

Thanks to the Olmecs, worship of the big cat spread widely over much of Mexico and Mesoamerica. Even today, in the month of May, in the remote village of Acatlán, in the state of Hidalgo, in central Mexico, Olmec descendants perform fertility dances to worship the jaguar.*[10]

Olmec beliefs advanced westward and as far north, some say, as the American Southwest, where the Mixtec, Zapotec, Toltec, and Izapa peoples adopted the big cat as their principal totem.

Apart from the Olmecs, no other civilization worshipped the jaguar as devoutly as the Maya. The jaguar sat at the apex of the Maya pantheon, commanding the sun, the night, the rain, and the underworld. It also governed the night sky. In fact, the Maya character for the night was the face of the sun embellished with the ears of a jaguar. The jaguar was so closely aligned with the night that its rosettes symbolized the stars spread in distinct patterns across the dark sky, and its eyes represented the gleaming moon.

The jaguar also played an important role in Maya pageantry. A jaguar pelt was the seat of Maya power, and "spreading the skin" was

---

* To build some of their structures, the Olmecs had to mine and transport massive amounts of stone over great distances. The jade and schist had to be brought in from the mountains. Obsidian found its way via established trade routes. As goods moved north, Olmec art, architecture, myths, rituals, mythologies, and belief systems traveled in the opposite direction, enabling many aspects of Olmec culture to reach far across Mesoamerica. Olmec culture also permeated much of Mexico. Saunders, *People of the Jaguar*, 55–57.

the term used when Maya warriors prepared for battle.[11] These same warriors wore the skin, teeth, and claws of the jaguar to intimidate their enemies. Yet the jaguar's power was not always related to violence. It was also associated with the miracle of life. Dennis Tedlock, a scholar of Maya culture, identified jaguar symbolism on a vase featuring scenes of women acting as midwives.

To underscore their authority, Maya rulers sat on thrones carved to resemble jaguars and wore jaguar skin robes. In fact, the pelts were considered so essential to the royals that jaguars were raised for the sole purpose of providing them with garments. The Mayas were also avid practitioners of ritual sacrifice. Jaguars—and humans—were killed to propitiate their gods. Some archaeologists maintain that the Maya priest-kings of Copán had grown so accustomed to blood sacrifice that they hunted the local population of jaguars to extinction.*

Around 850 B.C., not long after the Maya erected their sophisticated city-states, a civilization known as Chavín appeared, outside the traditional territory of the jaguar, taking its name from a small Peruvian village called Chavín de Huántar. Surrounded not by jungle but by the peaks of the snow-capped Andes, Chavín flourished, becoming the mother culture from which all subsequent Andean cultures were born.

The jaguar played an especially important role in the Chavín world. Chavín artists and artisans—like the Olmecs, who may have visited them via a land route or by sea—embellished the image of the jaguar. The Chavín believed that humans and animals could assume a variety of different forms, and their art captured this fluidity. Builders of the Old Temple at Chavín constructed an impressive plaza whose wall remained open to the east, in the direction of the rising sun and the Amazon rainforest. On the plaza's north side, artists carved fourteen jaguar-like figures with fangs, clawed hands, and serpents for hair.

---

* Maya architecture, like Olmec, celebrated the jaguar. In the ancient city of Tikal, Temple I commemorated Nu-Balam-Chakl, the "jaguar protector." Here the big cat was portrayed as a defender of the Maya priest-king, its claws reaching out to intimidate anyone who might dare to question his authority.

The Chavín civilization passed on their admiration for the jaguar to numerous other groups, including the Inca, who identified the big cat with their creator god, Viracocha. The Inca, however, adopted another big cat, the puma, as a symbol of their political power.

Far to the north, as Olmec dominance faded, the focus of civilization in Mexico shifted to the valley of Oaxaca in central-southern Mexico. Here the Zapotecs thrived for two thousand years (500 B.C. to A.D. 1500). Like the Olmecs, they, too, worshipped the jaguar. In fact, some anthropologists believe that the ancient capital site of Monte Albán, built by the Zapotecs in the sixth century B.C., was constructed to honor the jaguar god Cocijo.

In A.D. 650, in the Mexican highlands, the militaristic Toltecs, who for a time were governed by a leader named Four Jaguar, rose to power. Though its origins are a mystery, some scholars say the Toltecs erected the religious center of Teotihuacán, the largest city in the pre-Colombian Americas, with pyramid-temples that rivaled the largest Egyptian pyramids.* Warlike symbolism was an essential part of the Toltec worldview. In the Palace of the Jaguars, the most prominent mural depicts a jaguar blowing a conch shell that drips with blood. Some scholars believe the shell symbolized an extracted heart, which in turn represented the act of conquest.

As the Toltecs expanded their domain, their ideas spread, too. By A.D. 500, Toltec jaguar motifs had migrated across much of central and southern Mexico, the Yucatán, and the Guatemalan Highlands. In fact, Chichén Itzá, the expansive Maya site on Mexico's Yucatán Peninsula, is rife with diasporic Toltec influences and jaguar imagery.

Much more recently (approximately A.D. 1200), the arid central plateau of Mexico became the next locus of power in the New World, when the Aztecs—or more appropriately the Mexica—an aggressive, imperialistic, Nahuatl-speaking people, having conquered the Toltecs, rose to power. The Aztecs, Mexico's last pre-Colombian civilization, borrowed liberally from the civilizations that came before them.

The Aztecs believed that the jaguar commanded the rains, which

---

* Some scholars say the Totonac culture was responsible for building Teotihuacan.

nourished the arid central plateau that lay at the heart of their civilization. In an Aztec origin myth, one of the four primary elements, the "earth world," was known as Nahui Ocelotl (Four Jaguar), and the constellation Ursa Major (the Great Bear) was called the Great Jaguar. The Aztecs even devoted a calendar month—*ocelotl*—to the jaguar. Those born under its sign possessed the virtues associated with the big cat—courage, pride, and a willingness to do battle.

Armed conflict was an essential part of Aztec life, and no two groups were more revered than their two warrior clans, the elite Eagle Knights and the Jaguar Knights, whose responsibility it was to protect the Aztec capital of Tenochtitlán (today's Mexico City). The warriors of these two clans were so respected that, according to some scholars, the mountaintop temple of Malinalco, west of Tenochtitlán, was built for the sole purpose of honoring them.

Perhaps the singular characteristic of Aztec culture was what the archaeologist Nicholas Saunders called its "ideology of sacrifice."[12] In the feast of Tlacaxipehualiztli, or "the skinning of men," sacrificial victims were tied and tethered to a massive stone. When the Eagle and Jaguar-knights entered the room, armed with wooden swords, embedded with sharpened obsidian blades, they slashed at their captives. As blood dripped from the captives' wounds, trumpets sounded, announcing the horrible climax of the ceremony. The warriors released the prisoners from their tethers, then using knives cut the beating hearts from their chests and offered them to the sun.

CHAPTER 8

The Edge

IN GEORGE SCHALLER, ALAN RABINOWITZ HAD THE FINest mentor that the world of jaguar biology could offer. And in Howard Quigley, he found a lifelong friend. At the University of Tennessee in Knoxville, Rabinowitz and Quigley realized they were "kindred spirits." Rabinowitz hailed from New York City and Quigley from San Francisco. Quigley was laid-back and likable, while Rabinowitz had molded himself into the quintessential type A personality. He was competitive and driven, but like Quigley, he could be affable. But where Quigley collected friends easily, Rabinowitz, still embarrassed by his stutter, kept largely to himself.

Quigley was the more experienced of the two and had already spent considerable time in the woods. He had grown up camping and hunting with his outdoorsy father, and while an undergrad at Berkeley, he'd spent three summers doing black bear research in Yosemite. When he arrived in Tennessee and entered the wildlife program (Rabinowitz chose the ecology program), he had trapping and radiotelemetry skills, which he used in a black bear study in the Smoky Mountains. At Berkeley, he had done his undergraduate work under the inspirational A. Starker Leopold, whom he credited with being his motivation for entering the biological sciences.

Quigley and Rabinowitz began a tradition of getting together on Sundays for hamburgers, beer, and buoyant conversation. They shared their dreams of wandering the world while doing research as field biologists. One Sunday, as they settled in with a six-pack, Quigley told Rabinowitz about an article he'd read in the World Wildlife Fund (WWF) magazine. George Schaller was starting a WWF panda project in China. Quigley admitted that he thought he would be a good addition to Schaller's team. (In 1977 Quigley was in the last few months of his master's program; Rabinowitz had decided to go for his Ph.D.) But the idea of writing to someone of Schaller's stature intimidated him. Rabinowitz became Quigley's cheerleader, urging him to contact Schaller. Eventually, Quigley found the courage. After weeks passed, he gave up hope of hearing back. Then one day Schaller's wife Kay called and said her husband would be in touch when he returned from China. She explained further that Schaller would be calling not about the panda project but about the possibility of him joining Peter Crawshaw at the Miranda Ranch to help run the New York Zoological Society's jaguar project. Not long afterward Quigley was bound for Brazil.

On the morning of Quigley's first day in the field, he and Schaller were following a herd of white-lipped peccaries. Schaller was taking careful field notes, scribbling and sketching as he always did. When he saw that Quigley wasn't doing the same, he reminded him that recording his observations was part of the job. So Quigley took out his notebook and was writing diligently when he heard a sound. Turning around, he saw a jaguar in full stalk. The big cat was so intent on the peccaries, it hadn't noticed Quigley or Schaller. Flustered, Quigley did something that embarrassed him for the rest of his life. He was in the field with the world's preeminent mammologist, participating in his first jaguar study, and all he could do was to blurt out, "George, Jaguar!"[1]

To Schaller's credit, he never confronted Quigley about the incident, and Quigley was never sure if he had fallen a few rungs in his mentor's eyes. But Schaller had already dismissed the incident, and by the time Rabinowitz arrived in the Pantanal, eager to "suck out all the marrow of life," Quigley was a semi-seasoned researcher.

The Pantanal astonished Rabinowitz with its biodiversity (it is home to more than 4,700 plant and animal species) and its vibrant bio-intensity, so much so that he wrote, "Every part of the Pantanal was an assault on the senses." He could identify only a small percentage "of the more than 500 birds recorded in the region."[2] And he was thrilled by the Pantanal's beasts. "Every pool we passed and every stream we waded had multiple 'eyes' floating on the surface. Usually, those eyes were attached to bodies up to three meters (10 feet) long. . . . I marveled at each new encounter with caiman, anacondas, boa constrictors, and tarantulas on top of the hundreds of ticks, mosquitoes, and biting black flies. . . . The bugs were almost unbearable at times."

Under Quigley and Crawshaw's guidance, Rabinowitz learned the basics of field research. During those heady and challenging days, while keeping tabs on seven jaguars, the three researchers spent long days in the field, often with only a few hours of sleep. Exhibiting the kind of imperviousness to discomfort that became one of the hallmarks of his career Rabinowitz was a sponge, "champing at the bit" to spend every waking moment in the field to absorb as much as he could.[3]

"Imagine his excitement," Quigley said of Rabinowitz. "He'd been working mostly with raccoons in the Smokies and then he's in the Brazilian Pantanal tracking a thrilling apex predator."

Rabinowitz was a quick study. He learned telemetry, how to set wire cable foot snares, shoot dart guns and calculate the safest tranquilizer levels, how to fit and fasten a radio collar around a jaguar's neck, and how to triangulate a jaguar's position based on a signal. In Belize, because of the density of the forest, Rabinowitz did this almost entirely on foot, but in the spacious Pantanal, he tracked jaguars on horseback and by boat. He also learned, much to his mortification, that he was allergic to big cats. He vowed never to allow the symptoms—swollen and watery eyes, runny nose, and itchy skin—to prevent him from doing the work he loved.

All the while Rabinowitz was studying jaguar behavior. He learned that jaguars were fierce boundary keepers, using scrapes, scent rubs, scat deposits, and urine and anal gland sprays, to convey a variety of arcane messages. Thanks to their vomeronasal receptors, other cats were literally able to taste this information. But for Rabinowitz these complex signs were still like "hieroglyphics . . . pieces of which [he] could decipher but never understand in its entirety."[4]

Early on he shadowed Quigley and Crawshaw, but as he grew more comfortable, he ventured out on his own, walking the lonely dirt roads while tracking but never seeing the big cats. The most important thing he learned was to be a studious and curious observer. That was the key to field biology—to perform good methodical science, with a healthy dose of intuition mixed in, while still retaining the ability to appreciate the wonders of the natural world.

Rabinowitz flourished in the Pantanal, his experiences there "steeling [him] mentally and physically" for what he would ultimately encounter in Belize.[5]

IN THE WORLD OF WILDLIFE ECOLOGY, they say that large carnivore biologists often resemble the apex predators they study, and Rabinowitz was no exception. As a young man, he cultivated a fierce willpower and an abiding sense of his own physical indomitability. His longtime friend Jane Alexander, the actress, writer, and former director of the National Endowment for the Arts under President Bill Clinton, wrote that through an inexhaustible energy, he accomplished "more in a week than most of us do in a month."[6]

Another friend, Andy Sabin, described him as "totally and powerfully focused," believing that "only with strength [was] there discovery." Proud of his physique to the point of vanity, he even brought his barbells with him to Belize so he could continue his demanding daily workouts.[7] And when he exercised with his weights, the Maya gathered around, amused by his exertions and by his excessive grunting and groaning, repelled by the sight of the sweaty, animal-like hair on his chest. Why, they wondered, would a man work so hard when he wasn't working? The lives of the Maya were filled with wearying physical labor, and when the men weren't in the fields, tending to their corn plots or cutting and burning the forest, they lay in their hammocks, while the women spent long hours crushing and grinding corn. Their worldview taught them that their lives were expendable. Rabinowitz, on the other hand, was dedicated to a physical fitness regimen that he secretly hoped would allow him to defy death itself.

Not long after Guermo died, the wet season began. Though initially discouraged by what the incessant rain meant for his project, Rabinowitz used the time to construct a jaguar trap with a trapdoor and a small, protected platform at the back where he would tie a pig as bait. He hoped his ingenuity would pay off and that in the future he could rely on the trap instead of running jaguars with dogs through the bug- and snake-infested forest.

Rabinowitz also used the downtime to get to know the Maya families even better, especially the women and the children who had initially avoided the crazy *el hombre tigre* (the jaguar man) with the green eyes of a big cat. They began to trust him, gathering in his hut at all hours of the day to watch him write in his journal, read, and sleep, even to use the primitive toilet he'd installed. They confided in him, too, about how they had treated jaguars before he arrived. He was grateful for their candor but saddened by the regularity with which they were killing not just jaguars but all wild cats.

In the previous four years, they admitted to him that collectively they had killed nine jaguars, two pumas, and three ocelots. In a nearby community, where many still wore jaguar teeth and claws as talismans, the numbers were much the same: four jaguars, one puma, four ocelots, and a margay. He learned too that the Maya treated endangered scarlet macaws in the same fashion. They shot them on sight, ate them, then used the feathers to decorate themselves for ceremonies.

This information shook Rabinowitz to his core. He wondered how, given the frequency with which people were killing wild cats, he could possibly help the jaguars of Cockscomb. Would his efforts even matter in the end?

As the rains tapered and then ceased, Rabinowitz succeeded in capturing another male jaguar. He named the jaguar Chac, after the Maya god of rain.

Despite his uncertainty about the long-term benefits of his survey,

Rabinowitz was assembling the kind of data that Schaller and Carr expected. He amassed a collection of twenty-two fecal samples, which provided him with valuable information about the jaguar's choice of prey. Cockscomb, he discovered, functioned as a unique ecosystem. Because of the high density of prey, competing jaguars had the ability to coexist. Each one carved out its relatively small territory and, by identifying that territory, was able to avoid confrontations with other big cats. He also noted that the big cats, which knew of his presence long before he stumbled upon their tracks, never ambushed him.

Soon after capturing Chac, Rabinowitz trapped, collared, and released another jaguar. He estimated its age—with worn canines and a coat infested with ticks and the breathing holes of botflies—to be eight years old. He gave the jaguar the name Itzamná. The most important god in the Maya pantheon, Itzamná reigned over the heavens, and was often depicted as a toothless man with hollow cheeks.

Months later Rabinowitz went searching for Itzamná. For three days, the big cat's collar showed that he hadn't moved, so Rabinowitz called a local bush pilot, and together they searched for Itzamná's signal. Having found his location from the air, Rabinowitz and Cirillo, his new Maya tracker and assistant, set out on foot, slashing trails in the direction of the faint signal. "Vines pulled at our arms," Rabinowitz wrote, "and a dark, sulfurous mud gripped our boots. The area was unearthly, almost evil."[8] Cirillo seemed nervous, and unwilling to continue. But together the two of them found the jaguar's white skeleton picked clean. Rabinowitz wrote that the sight made the two of them "shiver" with fear. Devastated, he gathered up the bones, and when he returned to his shack, he reassembled the skeleton, using wire, and hanged it ghoulishly near his front door, as a warning to all spectators: Cockscomb's jaguars were sacred. The superstitious Maya began to speak of the presence of evil spirits and to wonder once again about the "crazy white man" living in their midst.

Not long after Itzamná's death, Susan Walker, Rabinowitz's girlfriend from the States, visited him. The first thing that greeted her was the jaguar's skeleton. Shaken at first by the sight, she could tell how thoroughly Rabinowitz had discarded any semblance of his former life. He spoke in truncated sentences, peppered with Creole phrases, and inhabited a small house stinking of recovered carcasses, drying skulls, dead scorpions and tarantulas, fer-de-lance heads, and jaguar fecal samples.

Though Walker arrived with news from home, she found Rabinowitz uninterested in the outside world. Consumed with Cockscomb, and Itzamná's death, all he wanted to talk about was his beloved jaguars. Another person might have feared that Rabinowitz had gone over to the dark side, a modern-day Kurtz, swapping the benighted Congo for Belize. But Walker, who had a bachelor's degree in anthropology, soon became fascinated with Rabinowitz's new existence. She embraced the jaguar project and the local Maya and immersed herself in the daily life of Cockscomb. Though she detested the ubiquitous mosquitoes, the burrowing botflies that embedded themselves in everything that moved, and the hellish heat, she loved the pace of life, even when downpours of rain kept her and the increasingly moody Rabinowitz housebound for days on end. Far from being a burden to Rabinowitz, she became an asset, venturing at night into the jungle, filled with alien sounds and the occasional poacher, to radio-track the big cats and gather data on their nocturnal wanderings. On rainy days, she tried to make sense of Rabinowitz's long-neglected paperwork. She discovered that the project, which Rabinowitz started eight months earlier, was over budget and would likely run out of money before the year was up.

Not long afterward Hassan sent word to Rabinowitz that he had captured a small female jaguar that he had been hired to kill for taking cattle but had had a change of heart and was keeping the animal in a cage. He wanted to know if Rabinowitz had any interest in trying to relocate her to the Cockscomb area. Rabinowitz named her Ixchel, wife of Itzamná, the Maya goddess of the moon. He released her where he had captured Itzamná, in an area of high prey density, hoping that his attempt at translocation would succeed.

Barely a week after bringing Ixchel to Cockscomb, Rabinowitz had to return to the States to persuade George Schaller and Chuck Carr to extend the study for another year. The good news was that Schaller and Carr, delighted with the research, were willing to prolong the project. The bad news came in the form of a telegram from Walker. Ixchel, she said, was dead.

When Rabinowitz returned, he learned from Walker, who had been remotely tracking Ixchel day and night, that the big cat had resumed her old ways and was once again killing cattle. Eager to collect the $300 reward placed on her head, a number of hunters had been searching for her. A caretaker from a local ranch eventually shot her. Tearfully, Walker reported that the caretaker had watched her when she was recording Ixchel's compass bearings. Rabinowitz had hoped that Ixchel might help him prove that the translocation of jaguars was a potential tool for dealing with problem cats. But Ixchel's homing instinct was too strong and her cattle-killing habits too deeply embedded. It was those habits that puzzled him. Why would one jaguar prey on cattle when a dozen others showed no interest? He conjectured that the cat had learned to feed on cattle as an adolescent and, like a young Dumpster-diving black bear, was unable to break the habit as an adult. Reluctantly he admitted that in order for ranchers to accept jaguars, they had to be allowed to shoot problem cats, though he acknowledged the arrangement could be ripe for abuse.

After Ixchel's death, Rabinowitz again had a crisis of faith. Cockscomb was losing its jaguars due to habitat destruction, revenge killings, and the long-standing habits of the local Maya. He wrote in *Jaguar* that the big cat was being eradicated from its historic range because it was "no match for . . . man" and that the death of the species was the "inevitable conclusion."[9]

Back in Belize, he drove himself unmercifully and neglected everything else in his life—Walker, his physical and mental health, and his relationships with the local people. In his first year at Cockscomb, he welcomed the local people into his crowded shack and ministered to their medical needs. But the more obsessed he became with studying and saving Cockscomb's jaguars, the less responsive he became to anything other than his mission. He grew impatient and lost his temper. At night, he slept fitfully and dreamed he was being stalked by a malevolent jaguar. He also became almost violently protective of Cockscomb, so much so that when he learned of credible threats to his life, he ignored them. It took a plane crash while performing an aerial survey of Cockscomb to reveal to him just how close to death he had come.

Weeks after the crash, he flew to the United States for a CAT scan and medical consultation. His doctors advised him to stay until his headaches and dizziness subsided. He refused to listen either to them or to Schaller or Carr, who, concerned about his health, encouraged him to "shut down the project."[10] He returned to Belize, as determined as ever to resume his survey. Time was precious, and there was too little of it for him to stand idly by and watch as Cockscomb was destroyed.

To the Maya, the cause of his torment was obvious. They believed that evil spirits were pursuing him. Rabinowitz dismissed their warnings, but sometimes, in the unbroken darkness of night, he felt like a frightened child.[11]

One day, not long after returning from New York, he and Walker were headed back to camp when they spotted Ah Puch near a creek. As they forded the creek in their truck, the big cat slunk down, as if ready to pounce. Rabinowitz hadn't seen Ah Puch since collaring him, and he noticed that the jaguar had grown gaunt. As he inched forward, Ah Puch melted into the forest. But when he stepped out of the truck, Rabinowitz could still feel the jaguar's presence.

Two days later he caught Ah Puch in a box trap. When he tried to sedate him with a syringe attached to a pole, the big cat reached through the bars of the trap and swiped at his face. Rabinowitz fell backward. Walker, who had been watching in horror, knew the tumble had saved him.

Rabinowitz and Walker waited for the drug to take effect, and when Ah Puch grew groggy, Walker carefully opened the trap's door. Rabinowitz reached in, grabbed the jaguar by his radio collar, and dragged him out. Still not entirely sedated, Ah Puch swatted at the air with his paws. Stunned by his own recklessness, Rabinowitz stepped back and waited for the sedative to take effect. When it did, he separated Ah Puch's upper and lower jaw and looked inside his mouth. He realized in a matter of seconds that Ah Puch, whose upper canines had snapped off at the gumline, had been in a state of agony. He could see the unprotected nerve endings dangling where the jaguar's teeth had once been. Heartsick, he knew the big cat had likely snapped off his teeth while biting in a frenzy at the trap's steel bars. How, he wondered, could he have inflicted so much pain on such a wondrous animal? The realization was almost more than he could bear. He had known that individual jaguars might be injured, but he had convinced himself that in the big picture he was helping to save a species. Now he felt no different than the thoughtless hunters who killed jaguars for their pelts or the cattlemen who shot them on sight. In his diary entry on November 19, he wrote, filled with self-loathing, that he had "interfered with nature's art" and caused "irreparable damage."[12]

After the incident with Ah Puch, neither Rabinowitz nor Walker could shake the feeling that their relationship was doomed. Together they made the painful decision that she should return to the United States. The local Maya, who had come to care about Walker, held Rabinowitz responsible. They believed he had driven her away with his moods and his blind loyalty to the jaguar.

Not long before she left, Rabinowitz and Walker trapped a young male jaguar, weighing barely eighty pounds. They named him Xaman Ek (pronounced *zamanek*) after the Maya god of the North Star. Rabinowitz wrote cathartically, "The raw beauty and energy of one of the greatest wild creatures on earth filled our hearts."[13]

Days later tragedy struck again. Someone had discovered a collared jaguar lying dead on a footpath. Rabinowitz knew it could only be Ah Puch. Still haunted by guilt, he felt rage rising inside him. As he walked away, he heard a Maya woman say, "He crazy, crazy bad."[14]

When he reached Ah Puch and knelt down, he noticed that the jaguar labored to breathe. Despite the large dose of penicillin he had given the cat days earlier, he smelled the stench of infection. As Walker and onlookers from camp watched, Ah Puch, possessed by some kind of wildness, lashed out with his paw. The swipe tore Rabinowitz's shirt, and the cat's claws cut open his shoulder. Then Ah Puch bit down on one of Rabinowitz's fingers. When he cried out for help, the onlookers could only stare on in disbelief. Left alone, he pried open Ah Puch's mouth and slipped his finger out before the cat's jaws clenched down again. With tears spilling from his eyes, he stood and lifted Ah Puch over his shoulder as if he were carrying a fallen soldier. As Tom Miller of *The Washington Post* would later write, Rabinowitz had become "a man dangerously close to the edge."[15]

Walker and the Maya watched Rabinowitz stumble down the path under the weight of the jaguar. When he reached his shack, he lay Ah Puch on the floor and sat beside him. Silently, Walker and the Maya crowded into the room. Rabinowitz ignored their presence until Ah Puch began to pant. Then he yelled for them to leave. "Get out," he screamed. "Get out, all of you."[*16] Unable to tolerate Ah Puch's suffering, he filled a syringe with a "massive dose of Ketaset" and injected it into the dying cat. He stood again, lifted the jaguar from the floor, and cradling him in his arms, carried him outside and set him in the back of the truck. When he reached the spot where he had first captured Ah Puch, he laid the jaguar in the grass, closed his eyes as if he were performing a sacrament, and asked Ah Puch for his forgiveness.

---

[*] Rabinowitz had a similar ourburst in Thailand. Devastated by a leopard's death, he reacted with anger and sorrow. There "was too much emotional intensity in the air," he wrote. The people who witnessed his display of emotion considered him to be *jy rawn*, literally "hot-hearted," or aggressive and impatient and voicing negative opinions. He "was never forgiven for the incident." Rabinowitz, *Chasing the Dragon's Tail*, 55, 100.

CHAPTER 9

He Who Rides a Tiger Cannot Dismount

FOR RABINOWITZ, AH PUCH'S DEATH WAS THE FINAL straw. He knew that his actions had precipitated the jaguar's demise, but he realized too that wild cats in and around Cockscomb were being killed every year, shot with utter nonchalance for revenge and for their pelts. In fact, the number of deaths he documented during his time in Cockscomb was overwhelming: ninety jaguars, fifteen pumas, seventeen ocelots, and six margays. He was convinced the actual numbers surpassed these.

In order to put an end to the killing, he understood he would need to protect Cockscomb. If not, cats would continue to die, and the wild basin would shrink as loggers encroached. For the first time since he arrived, he could look at the tree-shrouded hills and see ugly openings. Cockscomb, he feared, would be cut and cleared and bulldozed until it was nothing more than an "isolated forest patch."[1]

He'd also come to understand that Cockscomb's jaguars, and jaguars in general, needed room to survive. He estimated that twenty-five to thirty jaguars lived in Cockscomb's 154 square miles. Because of the plenitude of game, the basin could sustain a high density of big cats. If that prey base were diminished, its jaguars would suffer. So he established two goals, the most immediate of which was the preservation of the forest. Ultimately, he hoped to protect the "corridors" too, especially those along the creeks and rivers that allowed jaguars to rove in and out of Cockscomb.

As early as 1984, nearly twenty years before he would articulate his plan for the Jaguar Corridor, he understood that these paths, connecting critical areas of habitat, were the keys to saving the species. He saw that even Belize's most expansive protected areas, designated not long after independence, under the National Park System Act and the Wildlife Protection Act, were too small and isolated to maintain healthy populations of wide-ranging animals. In *Jaguar*, he wrote that this made the corridors "crucial for such ecological processes as dispersal and genetic flow."[2] He added, however, that before he could contemplate corridors, he needed to secure a "home ground," an unassailable tract of land, where jaguars could flourish. In the early 2000s, when he embarked on the Jaguar Corridor Initiative, he and his team would come to call these "home ground(s)," these core areas, Jaguar Conservation Units (JCUs).*

As Rabinowitz renewed his commitment to saving Cockscomb, his health declined, and his sense of his own physical and psychological indestructibility wavered as never before. He had always been proud of his physicality. According to Jane Alexander, it was something he hadn't "flaunted in a macho way." His vigor and athleticism were "just a part of who he was."[3] But now, as the headaches from the

---

* The concept was one Rabinowitz borrowed from his tiger days. In 1997, almost fifteen years after his Cockscomb experience, as the director of the WCS's big cat program, he convened a meeting in Washington, D.C., to address what he had come to call the "tiger dilemma." The group created a map of the tiger landscape and divided it into five bioregions. Then, within each bioregion, they pinpointed areas that had the habitat, prey, and protective status to sustain solid tiger populations. They called these Tiger Conservation Units (TCUs). After identifying the TCUs, they searched for corridors that had the potential to connect them, establishing the first-ever model for large-scale tiger conservation. Prior to that, tiger conservation had been blind to the bigger picture, focusing instead on small-scale, site-specific protection. Rabinowitz, *Indomitable Beast*, 120–21.

plane crash persisted and his "body was rotting away" from amoebic dysentery and intestinal hookworm, he implored a God he did not believe in to grant him the energy to "explore more wilderness" and protect the animals he loved. He longed to "feel strength permeate [his] entire being again."[4]

Psychologically, he struggled, too. He felt an "intense weariness" and confessed in his journals his growing ambivalence for Cockscomb, which he "hated, feared, and loved."[5] He worried he might never recapture his earlier idealism. He felt, too, that he had grown too close and become "too emotionally involved" with the jaguars.*[6]

When given the opportunity, however, to defend them, he accepted an invitation from the prime minister of Belize and his cabinet to deliver a speech on the state of Cockscomb's jaguars. He was still terrified of speaking in public, but he knew he could not reject the offer out of fear embarrassing himself.

For weeks, he practiced his speech, standing in front of a cracked piece of reflective glass that he used as a mirror. He detested the way his mouth contorted when he felt the urge to stutter, or stumbled uncertainly over a word, and he feared when in front of the dignitaries that he would so completely lose his nerve, he would once again be incapable of forming sentences. He later said he was so petrified of speaking in public that he'd rather walk through a pit of fer-de-lances.

Two days before he was scheduled to deliver his speech, he rose earlier than usual, shaved, took his blazer, shirt, and tie out of the closet, and folded them. He'd been humiliated into buying these respectable city clothes a year before, after attending a cocktail party in which a woman from Belize society remarked when she met him that she'd expected someone more distinguished looking.

When Rabinowitz arrived in Belize City later that day, the smell of the city's open sewers assaulted him, and he lamented the unfortunate habit of its residents who every morning dumped their "shit buckets" into the local harbor or the canals that traversed the city.[7]

---

* Schaller felt that "emotional involvement" with the animals one was studying was a prerequisite for being a good field biologist. Schaller, *Naturalist*, 21.

He checked into the Hotel Mopan, located in the commercial district, known widely for its old colonial charm and its convivial bar. He napped in an air-conditioned room, and late that afternoon he went to the bar to steel himself with scotch whiskeys.

The following day he picked up Chuck Carr at the airport. Carr had flown down to provide moral support but also to emphasize to Belize's leaders that, among all its global projects, protecting Cockscomb was a New York Zoological Society priority. That evening the two men retired to the hotel's bar to plot their strategy.

The next morning Rabinowitz once again woke early to shave and dress, and he and Carr walked down the street to the stately Government House. No sooner had they walked in than they were introduced to Belize's prime minister, George Cadle Price, a former seminarian who would soon be known as the "Father of the Nation." Price greeted them eagerly and surprised Rabinowitz by asking him to imitate the call of the jaguar.

"People talk about their calls," Price said, "but I have never heard them."[8]

Rabinowitz approximated the chilling sound of a jaguar grunting in the middle of the night, and Price roared with appreciation. Rabinowitz later wrote that it was the ideal "icebreaker."

Rabinowitz followed up his performance with an equally rousing slideshow in which he told his favorite stories. Despite his stutter, he was a natural storyteller, and his audience was electrified by his adventures. But then came the hard part. He had to convince the prime minister and his cabinet that Cockscomb needed to be protected both to save the jaguars and to safeguard it against the country's rapid deforestation. He prayed to the jaguar gods that his confidence wouldn't fail him.

Rabinowitz launched into the "economic" part of his talk. "I didn't let up on them," he later wrote, adding that he and Carr spoke with the "fervor of condemned men pleading for a stay of execution."[9]

After the first hour, when neither Prime Minister Price nor his cabinet members asked for him to conclude his remarks, he took their silence as an invitation to continue talking. Later he joked that "lunacy

[was] a crucial component of real-world conservation."¹⁰ He talked for another hour and then thanked everyone for their attention. He sat down to eat his cold meal and steady himself with a glass of wine. While eating, he watched as Prime Minster Price turned to his minister of natural resources and instructed him to protect Cockscomb.

The next evening Rabinowitz and Carr followed their talk with an encore performance, sponsored by the Belize Audubon Society. More than four hundred people showed up at the Bliss Institute theater overlooking the harbor. The event, in fact, was so well attended that people searched for standing room in the auditorium's gallery. Despite moments when his words caught in his throat, his mouth quivered, and his confidence wavered, he again delivered a fascinating presentation that the audience loved. For the coup de grâce, he asked Cirillo, his Maya assistant and friend, to join him on stage and demonstrate a jaguar's call. Cirillo's roar reverberated throughout the auditorium. Rabinowitz would write later that the experience and the reaction of the people, not only to Cirillo's jaguar cry but also to the presentation in general, was "overwhelming."¹¹

As elated as he was, he felt conflicted about the forces he was setting in motion. If his proposal to set aside Cockscomb as a reserve succeeded, the nine Maya families from the settlement, the people who had befriended him, would be removed from the place their families had occupied for centuries. Although they would be given commensurate land outside the reserve for their farms and milpas, they would resist leaving the only home they had ever known.

Rabinowitz had not yet resolved his feelings of having betrayed his friends. But just weeks before he was scheduled to shut down his project and leave Belize for New York, he learned that Cockscomb had been declared a reserve. Its official name was the Cockscomb Forest Reserve/Jaguar Preserve. The date was December 2, 1984. By the end of the month, the Maya would be gone.

AFTER RABINOWITZ LEFT BELIZE, Cockscomb underwent changes that surpassed his wildest hopes. The World Wildlife Fund made an investment, committing itself to protecting and maintaining the basin as a jaguar preserve over a five-year period. Its directors set up the reserve in partnership with the local Maya, who reluctantly exchanged their traditional land for the guarantee that only they could serve as rangers and park guides.

The directors made good on their promise and hired a father and son duo from the Maya community as the preserve's first wardens. One year later the Belize government declared the former timber camp a wildlife sanctuary, and Jaguar Cars of Canada donated $50,000 (and would later donate nearly $1 million to Cockscomb) to refurbish some of the camp's buildings, including Rabinowitz's former shack, as park structures. That year, 1986, 256 tourists visited Cockscomb, and the Belize Audubon Society assumed its management. The following year a Maya man became the sanctuary's first director.

In 1990 Belize's minister of natural resources expanded the sanctuary to include Cockscomb Basin's entire hundred thousand acres. The jaguars Rabinowitz had vowed to protect were inestimably safer. The Maya, initially bitter over being ejected from their land, came to accept their new home at the Maya Center. They held legal title to the land, and they embraced the opportunities that ecotourism offered. By 1994 they had a new school, a community center, and a health clinic, where a trained medical staff treated the skin wounds, parasites, and diseases that had been taking lives just ten years before.*

As Rabinowitz moved on, Cockscomb slowly faded from his memory. Shortly after leaving Belize, he met with George Schaller to discuss his future with the WCS. That future, according to Schaller, lay in Asia, which because of deforestation and the illegal trade in wildlife parts was losing its large mammals, especially its tigers and clouded leopards, at a frightening rate. Rabinowitz, now a WCS staff zoologist, eagerly followed his mentor's advice. In Borneo's dense rainforest and in the Huai Kha Khaeng Wildlife Sanctuary, at one

---

* Today Omar Figueroa, a jaguar researcher who studied the big cat in the Maya Forest Corridor, serves as Belize's minister for fisheries, forestry, the environment, and sustainable development.

thousand square miles the most intact forest in Thailand, he spent months tracking clouded leopards, "one of the most elusive cats in the world," and tigers. Then, having pushed himself to his physical limits, he returned to the United States to recover.[12]

Back in New York, he again sought Schaller's counsel. Some, like Peter Matthiessen who traveled with Schaller through the wilds of the Dolpo region of Nepal, considered him exacting and distant. Rabinowitz knew that the dignified but tough-as-nails Schaller was simply misunderstood. He was a resolute man who preferred being alone on a "snow-covered hilltop in the Himalayas to a crowded office in New York."[13]

Schaller hoped his protégé would return to Borneo, but Rabinowitz felt that he had left his work in Huai Kha Khaeng Wildlife Sanctuary unfinished. While meeting with Schaller in his nondescript office at the WCS headquarters in the Bronx, Rabinowitz tried to communicate his sense that there was still significant work to be done in Thailand. Schaller was concerned that he was taking on a project that was simply too ambitious. Rather than dealing with a single species, he would be embarking on an effort that encompassed a vast, complex, and imperiled sanctuary, as well as its wildlife. But Rabinowitz was insistent, and while attempting to win over Schaller, he hit upon the idea that he would use to help save jaguars across the Americas. He would concentrate on tigers and leopards, not only because they were two charismatic animals that the government cared about, but because tigers and leopards were umbrella species; if they were thriving, the entire ecosystem was flourishing. If they were suffering, it was a "warning signal for wildlife management and conservation."[14]

As Rabinowitz described his plans to concentrate on the big cats, Schaller sat calmly in the chair he rarely used because he spent so much time in the field. Although no scientist in the world knew more about tigers than he did, Schaller gave Rabinowitz his full attention. In the early 1980s, Schaller had conducted the first scientific study of Bengal tigers in India and wrote *The Deer and the Tiger*, a book that biologists considered to be the definitive account of tiger behavior. Schaller had also mentored other young and idealistic zoologists, including Dian Fossey and Jane Goodall. He could have pulled rank on Rabinowitz, and surely he was tempted to do just that. But when Rabinowitz finished making his argument for returning to Thailand, he instead stared intently at his student and said, "Put together a budget and a proposal for spending at least two years there." Then, knowing how headstrong Rabinowitz could be, he added, "Don't try to do too much at first."[15]

Three months later Rabinowitz was bound for Thailand, where he spent the next two years of his life. "Spirits, monks, tigers, and leopards," he wrote. "I had no idea what to expect."[16]

CHAPTER 10

El Tigre

AT NOON ALEJANDRO AZOFEIFA AND HIS SMALL TEAM OF park rangers walked out of the forest. A scarlet macaw announced their arrival, and the white hawk that had been perched like a sentinel on the pole where the Costa Rican flag drooped in the still air turned its head slowly and fastened its serpent eyes on the men. Azofeifa wore a blue polka dot bandanna, a khaki shirt, and military fatigues. He moved purposefully, like a man accustomed to finding his way even in the deep jungle. A machete scabbard swung from his waist, and I noticed that he and one of his junior rangers were armed. Word was that in addition to their survey responsibilities, Azofeifa and his team were doing potentially dangerous patrol work, reporting and sometimes chasing poachers, miners, and narco-traffickers who were using the park's dense forests and empty beaches to move drugs north through Nicaragua and into Honduras.[*] This was Corcovado National Park's dark side, one that the Costa Rican government and the ecotourism industry had struggled to keep quiet.

When Azofeifa reached the steps of the modest, clapboard building

---

[*] Eduardo Carrillo, Costa Rica's jaguar expert, experienced death threats and relocated his research to Guanacaste National Park.

that served as the rangers' barracks, he swung his pack from his back and dropped it onto the ground, happy to be rid of it. I was sitting on the covered porch at the Sirena Ranger Station (Estación Biológica Sirena) trying to escape the merciless midday sun. The station, which had been carved out of the forest, had an expansive, close-cropped lawn and neat, well-maintained buildings connected by covered boardwalks. A handful of European hikers and their guides sat on the porch, too, taking a lunch break when the sun was at its hottest.

When one of the guides shouted to Azofeifa, "Hey, Champeon!" the others chimed in, "Champeon!" The world of wildlife ecology may be an insular one, but Alejandro Azofeifa clearly had achieved celebrity status in Costa Rica's Osa Peninsula.

I watched Azofeifa as he sat down on the outside steps, wiped the sweat from his face with a bandanna, slipped off his Wellington boots, and chugged the water in his canteen. Only then did he smile, wide and friendly, "*Pura vida, muchachos.*" Then he stood and walked into the ranger house.

The jaguar and Azofeifa had brought me to the Osa Peninsula and Corcovado National Park, where the largest remaining tract of Pacific Tropical Wet Forest in Central America still stands. I hoped to work with Azofeifa's ranger team, which was setting camera traps throughout the park in an attempt to assess the health of its jaguar population. My most fanciful hope was to lay my eyes on a jaguar and maybe even on Macho Uno, Corcovado's most famous big cat. Half an hour later Azofeifa emerged, looking showered and refreshed and entirely different from the person I had seen walk out of the forest. He wore the requisite Costa Rican goatee, neatly trimmed and flecked with gray, and he was small, especially in relation to the person I had pictured. I'd heard so many stories about him that I'd expected a giant, a deep-voiced man as broad-shouldered as Alan Rabinowitz and as powerful as a jaguar.

Then the banter began, and I picked up what I could. The word I kept hearing was *el tigre*. And each time I did, I felt an electric jolt. *El tigre*, the name, is used in Mexico and much of Central America to refer to the jaguar. In Brazil, the jaguar is known as *onca*. In Suriname, it is called *penitigri*. The word *jaguar*, in turn, is derived from the Tupí-Guaraní word *yaguara* or *yaguareté*, meaning, among other things, "beast that kills its prey with a single bound" or "that which seizes" or "eater of us."[*][1]

Though jaguars are ignorant of nationhood, Corcovado's big cats may be some of Latin America's safest—though that wasn't always the case. When Rabinowitz first visited the country in the autumn of 1984, he lamented the fact that although Costa Rica had set aside parks and reserves, few truly wild places for jaguars existed. By the 1990s, 80 percent of the country's land had been heavily logged. Alarmed conservationists, including Álvaro Ugalde, dubbed Costa Rica's "conservation crusader," fought back, proposing a stringent new model of protection that resulted in 28 percent of Costa Rica's territory being set aside for environmental protection. Then the country passed a far-reaching law prohibiting the cutting of trees without a detailed management plan and initiated a program, funded through a national tax on gasoline, to pay landowners to maintain their forests and plant new trees. Today Costa Rica is reaping the benefits. Its forest cover has more than doubled, and it is well on the way to achieving its goal: trees blanketing 60 percent of the land by 2030. With the trees, the wildlife, and especially the jaguars, have returned.

Despite my burning desire to see a jaguar in Corcovado, it didn't take me long to realize how preposterous that idea really was. Jaguars are almost unseeable, somewhere close to invisible. With their mottled coat, they are masters of camouflage, able to meld into their surroundings—a feat known in the scientific world as crypsis—to avoid being detected even when they're just a few feet from their prey. Eduardo Carrillo, who has been studying jaguars for forty years and was in charge of the WCS's jaguar program for Central America from 1998 to 2004, has always marveled at their ability to remain undetected. "I can be standing right next to one," he said, "and I know it

---

[*] "The etymology and translation of the term yaguara has long confounded scholars," writes Wilcox. Walter William Skeat "reported four translations of this term: 'that which seizes,' 'the eater,' 'the dog,' and 'eater of us.'" Wilcox, "Encountering El Tigre," 81.

because I've picked up the signal from its radio collar, and, still, I may never see it."[2] He added that sometimes he relied on the screams of monkeys to alert him to the presence of a jaguar. He also trusted his nose. There were occasions when he would smell one's pungent odor without ever seeing it. Carrillo's graduate students were all too familiar with the cat's stealth. Roberto Salom-Pérez, now the Latin America director of Panthera, who, while earning his master's degree at the University of Costa Rica, studied under Carrillo (as so many of Costa Rica's jaguar biologists have), never once laid his eyes on a jaguar in the nine months he was working in the field in the Osa Peninsula.

Although there are places in Costa Rica where *el tigre* is barely tolerated and is even reviled to the point of cruelty, in Corcovado National Park, and in parts of the Osa Peninsula, it is often worshipped. Jaguar imagery is everywhere. Rivers, peaks, restaurants, art galleries, and lodges are named after it. In the pretty Puerto Jiménez Hotel, where my wife and daughter were staying while I was in Corcovado, guests are greeted by a colorful jaguar mural that adorns a large wall. When I first encountered it, I was surprised by the veneration. But the more I learned, the more I understood that jaguar mythology, which had once reigned over much of Mesoamerica, persists in small pockets across the jaguar range. Historically, the big cat struck a resonant note with people throughout the New World, and the mythology it inspired is as rich as any animal ever to walk the Earth. In some places where this mythology has lapsed, reverence for the jaguar has been resurrected in the name of ecotourism.

Solitary and secretive, jaguars often seem to be more ghost than animate animal. In the jungle, the jaguar tenses when it feels the touch of the sun because over a million years of evolution have taught it to creep among the shadows. Those who have been lucky enough to actually see one have been awestruck. Eduardo Carrillo calls it the most impressive piece of wildness he has ever laid his eyes on. In Corcovado, the puma, another formidable cat, is thriving; Carrillo has seen many over the years. Spotting a puma is a riveting experience, but seeing a jaguar is on a different scale altogether. In an interview with Natalie Angier of *The New York Times*, Carrillo, who has been fortunate enough to see dozens of jaguars in the wild, exclaimed, "It is like a miracle or a dream, the most exciting thing you can imagine."[3] He adds, "Every time I see one, my heart grows. I know that I am in the presence of something wonderful and feel close to God."

Many longtime biologists and rangers like Champeon who have wandered rainforests for decades, however, have never gotten more than a split-second glimpse of a big cat slipping away, or crouched so low to the ground it could have been a snake. But that split second, when the image of the big, beautiful cat was stamped indelibly on their brains, left them yearning for more.

This is the way it was with Champeon, who began panning for gold prior to the creation of Corcovado National Park. The establishment of Corcovado was spearheaded by the biologist Álvaro Ugalde, also the co-founder of Costa Rica's park system. He had urged the government to negotiate a complicated land swap for what would become Corcovado. In 1995 the park was still home to a collection of rough men, who lived outside the law. Peasant squatters called *precaristas*, poachers, timber harvesters, crocodile hide hunters, pirates, fugitives, and murderers populated Corcovado's forests. They hunted wild game for food and drank *guaro*, the cheap domestic rum that was sold in the tiny, makeshift *pulperías*. A miner addicted to the lure of gold, Champeon joined their ranks.

In 1995, twenty years after the creation of the park, he finally gave up gold mining and promised himself, like a lifelong drinker going cold turkey, he would never again pick up a pan. In late December 1994 he shot his last white-lipped peccary for food and then just weeks later began working for the National Institute of Biodiversity, whose plan it was to turn miners, hunters, and squatter farmers into allies, benefiting from their extraordinary familiarity with the forest. And no one knew Corcovado better than Champeon. The institute sent him to school to study taxonomy. Initially beetles were his specialty.

Exactly how Azofeifa became the champion of the jaguar is a bit nebulous. The simple story is that one night while seated by a fire along the park's Rincón River, where he was collecting insects, he was circled by a jaguar. He heard the bellows-like grunting of the big cat, hoarse, breathy, and primal. For some, it might have been a terrifying experience. But Champeon was thrilled. He could feel his heart pounding and hesitated to twitch for fear of scaring the jaguar off. Not long afterward he gave up beetles and made the decision to devote his life to protecting the jaguar and the rainforest it roamed. His story is surely less lucid and linear than that, but its essence is real and has been confirmed by Champeon himself. *El tigre* changed his life.

CHAPTER 11

The Magic

WHEN THE REPARTEE BETWEEN CHAMPEON AND THE guides ended, I walked over to introduce myself. In my best Spanish, I told him who I was and why I was there. When I finished, I realized he was puzzled, and I wondered what I'd failed to convey. Just then one of the other rangers intervened.

"My name is Dani," he said in crisp English. "Are you the writer who is here to follow us?"

Sí, I said. *Escritor.*

Champeon gave a curt nod of recognition.

Then Dani explained to me that they were nearing the end of their survey. They had spent the last two weeks wading through the jungle, sleeping in hammocks, and setting up camera traps as part of their JaguarOsa Project, their goal to observe populations of species—jaguars, white-lipped peccaries, and Baird's tapirs—that were reliable indicators of the park's health. Camera traps are similar to the trail cams that many hunters use to keep tabs on deer, elk, and bear. The cases are painted in camouflage and attached to trees and are activated automatically by heat and motion. Dani added that Champeon, who was part of JaguarOsa from the very start, was especially proud of his role in the study and in the scholarly papers JaguarOsa had produced.

CRESTFALLEN I HAD MISSED THE opportunity to travel at length with Champeon, I realized for the first time that my plan to follow rangers and biologists doing fieldwork throughout Central and South America might amount to a logistical nightmare, especially during Covid. But Dani assured me there were more camera traps to set and check and that I would be welcome to join their efforts. He added that they were finished for the day but planned to head into the forest again the following morning. So I shouldered my pack and in the searing morning sun took a short trail to the beach, Playa Sirena, which looked out over the vastness of the Pacific Ocean. Overhead black vultures floated in the thermals. Far up the beach, I came upon a mudhole and stopped to examine fresh peccary tracks. The air, even with an ocean breeze, reeked with the acrid odor of musk and urine.

White-lipped peccaries, *chancho de monte*, though infamously ornery, are a favorite jaguar prey, and like the jaguar, they, too, are a sign of a healthy and functioning ecosystem. According to Jaguar-Osa camera trap data, Corcovado is home to between four and seven "mega herds" of white-lipped peccaries, numbering 150 or more. However, in general, white-lipped peccaries are facing declines throughout the Neotropics and especially in Costa Rica, where they have been eliminated from about 89 percent of their historical range. Corcovado National Park is their last remaining refuge.

But that hasn't always been the case. In their surveys of Corcovado in the early 2000s, Eduardo Carrillo and Roberto Salom-Pérez documented an alarming drop-off in their population due to unchecked hunting at the edges of the park. Carrillo noted that hunters were using AK-47s with which they could decimate a herd in mere minutes. During that same period, he also noted a significant decline in the jaguar population. The solution, he said, was an obvious one. In order to protect the long-term viability of the jaguar, and the long-term well-being of the park, prey populations inside and outside the park needed to be protected. Apart from being plain old tasty food, packed with a high caloric payoff, peccaries are what might be called the jungle's gardeners. Their rooting helps to maintain the biotic diversity of the forest floor. They are also habitual wanderers—and poopers—that spread seeds as they roam.

I continued up the beach until I remembered what my friend Mike Boston, a legendary Corcovado guide, had told me one night, in his lilting Irish voice, as we sipped beers from our porch at Cabinas Jiménez overlooking the pretty Golfo Dulce (Sweet Gulf). Perhaps the most frightened he'd ever been in his two-plus decades of guiding in Corcovado was when a large herd of white-lipped peccaries, moving like a single animal, surrounded him and a friend. The peccaries were barking, rattling their tusks, and clattering their teeth so viciously that Mike and his friend climbed nearby trees and didn't dare to descend until they were certain the herd had moved on.

Leaving the beach, I entered the jungle, as a soft yellow light wafted through the trees. One hundred feet in, I spotted a magnificently iridescent morpho butterfly, one of the largest and most colorful butterflies in the world. Like so many other species suffering from habitat fragmentation, the morpho had found a home in Corcovado. I followed it as it flapped its beautiful metallic wings, moving languidly in and out of a shaft of light. I hadn't been walking long when I ran into Jason, Champeon's newest ranger, a snake enthusiast, who was diligently examining the leaf litter alongside the trail. Together we set off down the trail, occasionally leaving the safety of the path to investigate the vegetation. I moved slowly and attentively, especially careful of my footing, and watched Jason as I walked. His eyes were always roaming.

Corcovado National Park is full of deadly creatures. At or near the top of that list is the snake for which Jason was searching, the well-camouflaged and highly venomous *terciopelo*, the fer-de-lance, the very species that had killed Guermo in Belize. The fer-de-lance kills more people in Central and South America than any other snake. If angered, it is fast and extremely aggressive. It is also equipped with inch-long fangs, or "spear points" that function like hypodermic needles and are sharp enough to penetrate leather. What's more, it is capable of opening its jaws to a terrifying 180 degrees, leaping upward to deliver a strike, and shooting its lethal venom several feet through the air. More than 90 percent of those struck receive a significant dose. Unless antivenin is administered, death often occurs in less than four hours and is accompanied by swelling, blisters, and blood literally oozing from the pores of a victim's skin before hemorrhagic shock happens.

While Rabinowitz was studying jaguars in Belize, he came to terms with the lethal dangers of the jungle. But the one animal he loathed, with a hatred born of fear, was the fer-de-lance. Later in his career, when he was struggling to save Myanmar's tiger population, he developed a similar paranoia of its killer snakes, the cobra, krait, and Russell's viper. But in Belize, this passionate animal lover would kill a fer-de-lance snake without a second thought. In one scene from his book *Jaguar*, he swings his machete with all his might and slices a fer-de-lance in half. Its entrails spill across the ground, but the head keeps coming toward him, jaws open, as if in a horror movie.

As Jason and I were searching in a streambed, he for snakes, and I for jaguar tracks, we both heard the faintest snap of a breaking stick. That's when I remembered what Erik Olson, an associate professor of biology and natural resources at Northland College and co-supervisor of the JaguarOsa Project, had told me about being in a forest like Corcovado. "You can be assured, you're being surveilled," he said. "But that's part of the magic."

I could feel my body tense. Crouching like prey in the middle of a creek bed in a jungle that held jaguars—and also pumas—can be utterly disconcerting. According to Doug Peacock, the grizzly expert and writer, those conflicting emotions—exhilaration and terror—were exactly what I was supposed to be feeling. Edward Abbey wrote that his friend Peacock insisted that the only places that qualified as a real wilderness were those "where one enjoys the opportunity of being attacked by a dangerous wild animal." Abbey added, "A wild place without dangers is an absurdity."[1] In his classic *Coming into the Country*, John McPhee wrote similarly: "The sight of the [grizzly] bear stirred me like nothing else the country could contain. What mattered was not so much the bear himself as what the bear implied. . . . He implied a world. He was an affirmation to the rest of the earth that this kind of place was extant."[2]

In a world where genuinely wild places are becoming harder to find, the Osa Peninsula and Corcovado National Park are indeed exceptions. Lying just north of the Panama border, the Osa is a seven-hundred-square-mile protuberance. Surrounded on three sides by ocean, it looks remarkably like a jaguar's paw. It has been described as Costa Rica's Alaska, isolated, hard to reach, and untamed.

In the 1930s it was the epicenter of a legendary gold rush. Hardened miners, *oreros*, drenched in sweat and assaulted by rain, ticks, mosquitoes, and sand flies that carried the dreaded leishmania —called *papalomoyo*, or miner's leprosy, that ate away a man's face— panned for gold so pure and fine that jewelers all over the world clamored for *oro Madrigal*, Madrigal River gold. As recently as the 1990s, gold flakes trickled down streams, and miners still dreamed of striking it rich.

In addition to gold, Corcovado National Park harbors 2.5 percent of the Earth's life-forms. It is the Osa's main attraction, one of the world's biological hotspots, and is widely considered the crown jewel of Costa Rica's national parks and biological reserves. Five species of wild cat inhabit its forests, and depending upon one's source, it is also home to somewhere between eight and fourteen jaguars. In many ways, these jaguars are the new gold, attracting tourists from all over the world. Although the chance of actually seeing one is next to nothing, many travelers are inspired by the opportunity just to walk in a forest that still contains jaguars.

Decades ago Alan Rabinowitz lamented what ecologists call the "empty forest syndrome."[3] One of the problems with conservation, in his opinion, was that governments and organizations across the world created and protected beautiful parks, preserves, and chunks of land, lacking in key species, most especially carnivores, like the jaguar. Half a century before Rabinowitz, Aldo Leopold wrote similarly. He imagined life as a "biotic mechanism" or "biotic pyramid" with soil at the bottom and carnivores at the top. He knew that humans had the temperament and the means to do "unprecedented violence" to this pyramid, admitting that he, too, at one time, pursued predators with a lethal lust.[4]

But Leopold had undergone his own evolution with regard to apex predators. In 1933 he accepted a position as professor of game management in the agricultural economics department at the University of Wisconsin. It was the first time in the United States that anyone had ever been appointed a professor of wildlife management. He used the opportunity to articulate and disseminate his unique land ethic, which among other things defended the right of species— even those that were reviled and persecuted—to exist. In 1936, in an article called "Proposal for a Conservation Inventory of Threatened Species," he again upheld the role of predators in the natural world, as well as their innate worth. His visionary notion was that if we could change the way we treated apex predators, we could change the way we treated nature itself. At the time, grizzly bears, wolves, coyotes, mountain lions, and the occasional jaguar were being shot, trapped, mutilated, and poisoned in the United States, almost out of existence.

In his essay "The Green Lagoons," Leopold lamented the eventual disappearance of *el tigre*, including the loss of wildness that its absence implied. He wrote mournfully that "a glory [had] departed from the green lagoons." The delta of the Colorado that he had so loved had "been made safe for cows."[5] That essay had a profound effect on Rabinowitz, who wrote in the pages of *Indomitable Beast*, "I knew what Leopold was talking about. . . . After years in the jungle, I knew when jaguar habitat no longer had jaguars. . . . When the top predator no longer walked the land, no longer vocalized in the night . . . the energy of the forest had changed."[6]

Scientists had discovered that when apex predators were removed from the environment, an ecological process called trophic downgrading could occur—a thorough disruption of a natural system that happened when an animal at the top of the food chain was eliminated. As a conservationist, Rabinowitz understood the paramount importance of saving the jaguar, a leading "keystone" species that played an oversize regulatory role in the environment. In other words, the jaguar was the proverbial canary in a coal mine.

CROUCHED IN THE CREEK BED, Jason and I eventually laughed out loud when we realized what our addled imaginations had conjured. The jaguar we had imagined was nothing more than a group of curious and agile spider monkeys, coming down from the canopy and moving with their almost infallible sense of balance to investigate us.

Relieved, we both took a deep breath, watched the monkeys for a while, then spent the rest of the afternoon wandering Sirena's trails and exploring other creek beds. We never saw a jaguar track, or a fer-de-lance, but we did see plenty of wildlife—more spider monkeys, a group of agitated howler monkeys, coatis, kingfishers, short-billed pigeons calling "who cooks for you" from the forest canopy, a yellow-beaked great curassow, a venomous coral snake, and an anteater. We were even lucky enough to spot the back end of a tapir crashing through forest.

Late that afternoon, when we returned to the station, I noticed that a handful of tourists had gathered on the porch. When I asked one of the guides what the commotion was about, he said that someone had spotted a puma right there on the grounds, on the cut grass, a mere fifty yards from where we were standing.

After dinner I returned to the porch. The scarlet macaws had settled in, and the howler monkeys had ceased their maniacal grunting. Shortly afterward the skies over Corcovado turned black. For a moment everything was perfectly still, and then the rain came with a ferocious roar. Downpours are common on the Osa Peninsula: rains spill out of the mountains, turning rivers into raging torrents full of uprooted trees and debris, serving as a reminder that here, in this wild corner of Costa Rica, the elements still rule. But the rain is also a gift. It helps to make the Osa one of the most biologically intense places on the planet.

Two hours later the storm finally relented, and when it did, the forest came alive with the hum, clicking, whistling, and general racket of untold insects and amphibians. Then the sky cleared, and shards of starlight lit the ground.

CHAPTER 12

The Mother Lode

I WOKE HOURS LATER ON THE PORCH TO AN EARLY MORNing mist that filled the air and clung to the forest canopy. My watch said four a.m. as I walked from the porch back to my bunk to arrange my gear for the day, careful not to wake the few visitors. Because of Covid, 2020 and 2021 were especially bad years for tourism in Costa Rica. Corcovado National Park, in particular, witnessed a precipitous drop in tourism and a commensurate surge in illegal activity. Jobs in its once vibrant ecotourism economy had dried up, and people, driven by circumstance and the high price of gold, were again placer mining, rerouting creeks into sluice boxes, and sometimes using poisonous quicksilver mercury to extract the gold from the slurry. They were also poaching "bush meat" for food. In 2020 park rangers made thirty-four arrests. But because of Corcovado's size and its remoteness, some miners undoubtedly escaped notice.*

At breakfast, two hours later, Dani informed me that the ranger team would be checking a handful of cameras that day and then

---

* Juan Carlos Cruz, a jaguar researcher with NAMA Conservation in Costa Rica, lamented the conditions that drove disheartened people to commit environmental crimes. "We have a saying," he said. "Nobody does conservation with hunger."

would return to the station to write reports and to examine the camera cards they had collected. Having talked with Olson, the co-head of the project, before arriving at the park, I knew that the JaguarOsa team had been encouraged by the data it had amassed. "What we're seeing," Olson told me, "suggests that there's a lot of hope for the wildlife community of Corcovado National Park."

Sampling animal populations with camera traps was first used on tigers in India, but it has become increasingly popular, particularly for monitoring elusive species like the jaguar. According to Olson, the data from the JaguarOsa Project was shedding new light on jaguars and their habits, and that data was enabling him and his team to establish long-term conservation strategies for the park. Two of the most important revelations of the JaguarOsa Project were that jaguar populations in the Osa Peninsula had increased to a total of twelve (ten males and two females) since 2013, and that big cats, in a rich and diverse place like Corcovado, could grow considerably older in the wild than previously suspected.*

Two thousand seventeen was the year when, thanks to the fortuitous appearance of some forgotten photographic evidence, Olson became convinced that biologists had been wrong about the lifespan of jaguars outside captivity. Late that year, at the picturesque Northland College campus in Ashland, Wisconsin, on the shores of the windswept Lake Superior, Olson, who spends part of the year track-

---

* Carillo and jaguar researchers from NAMA Conservation dispute JaguarOsa's results, citing methodological inadequacies. "In the case of Corcovado, if we don't act quickly, the deterioration of wildlife populations could be irreversible." Eduardo Carrillo Jiménez, "El parque Nacional Corcovado agoniza . . . de nuevo," *Semanario Universidad*, July 20, 2021. Carillo maintains that Corcovado's jaguars are failing because of a scarcity of white-lipped peccaries, while the puma population is growing because their favorite food, collared peccaries, is burgeoning. In my interview with him, he told me that he has spent much of his professional life in Corcovado. "The Osa breaks my heart," he added. In a June 2022 piece for *Mongabay*, Maxwell Radwin described the disagreement: "The extreme polarization over what's going on in Corcovado National Park has led to accusations of corruption, negligence, media manipulation, fights for control of the area's management, and who does and doesn't receive funds from international donors." The accusations have been denied and have not been proven. Radwin, "Miners, Drug Traffickers and Loggers."

ing wolves in the cold, sprawling northern forests of the Upper Midwest, received a package from his Costa Rican counterpart Guido Saborío, who headed up Costa Rica's National System of Conservation Areas (SINAC). For two and a half years, Olson and Saborío had been running the JaguarOsa Project, which was funded, in part, by the college.

The package that Saborío sent contained photos of a single jaguar taken by a graduate student in biology who had been working in Corcovado National Park in 2008. Saborío included the student's master's thesis, but he was especially eager for Olson to see the photos and wondered if Olson would come to the same conclusion he had.

When Olson studied the images, concentrating on the jaguar's pattern of rosettes, he was nearly dumbstruck; they looked identical to Macho Uno's. Macho Uno, Male Number One, was the first jaguar they had captured on film when they began their camera trapping operation in March 2015. Macho Uno had paraded assertively in front of the camera, showing off the distinctive rosette patterns on both sides of his muscular body. Two months later they caught him on camera again.[1]

Olson studied the photos and even brought in some of his students to verify his findings. But each time a voice in his head shouted *Impossible!* A jaguar's design was its fingerprint. No two big cats had the same markings. Olson kept at it until he was certain beyond a shadow of a doubt that the rosettes were exactly the same. Then he knew he was looking at Macho Uno, albeit a younger, smaller, less developed version, so undeveloped, in fact, that the graduate student had mistakenly classified the jaguar as a female.

Olson, Saborío, and their team threw themselves into the research, consulting papers, reports, and social media posts about the lifespan of jaguars, all of which confirmed their exciting assumption: Macho Uno could be the oldest known jaguar ever to live in the wild. In their research, they discovered another captivating piece of information. Based on the biologist's findings in 2008, they realized that Macho Uno was very much a homebody. Jaguars are roamers, especially male jaguars. Territories are flexible and change depending on a host of factors, including the availability of prey. But Macho Uno seemed to have no need to wander. He was especially fond of the Corcovado Lagoon area, the *bajura*, the lush and remote lowland that lies in the dead center of the park. It had been his primary territory for the entirety of his life.

In 2018 the team captured photos of four male jaguars frequenting the same area: Macho Uno; a big, powerful male they named El Trotamundo, the "Wanderer"; a new jaguar they called Espejo, "Mirror"; and another jaguar they'd never seen before, a large, unnamed male. How, they wondered, could an old jaguar like Macho Uno compete with younger, stronger, and more aggressive males?

The following year they got their answer. Macho Uno was exhibiting what biologists called "extraterritorial movements." He was moving far afield, entering new areas where they'd never seen him before. When Macho Uno was captured on camera outside the park in the Golfo Dulce Forest Reserve, the team had reason to worry. In the reserve area, Macho Uno would not enjoy the protections of Corcovado.

By 2020, Macho Uno had disappeared. No one in the park or the reserve had seen him on their camera traps. Olson and his team had all but given him up for dead. But then in 2021, he reappeared on one of JaguarOsa's thirty-eight cameras. He was alive, in the twilight of his life.

AS WE WERE FINISHING UP BREAKFAST, Dani admitted that he had never seen a jaguar in the flesh. He said that seeing a photo or video of one, or the track of a large round pad pressed into the mud, was likely as close as he'd ever get to the real thing in Corcovado. He talked about Santa Rosa National Park in the northwest of the country, and Tortuguero National Park in the northeast. Seeing a jaguar was almost guaranteed there, especially during the sea turtle nesting season, when jaguars prowled the beach in search of sea turtles laying their eggs.

Prior to arriving at Sirena, I had discussed both places with

Roberto Salom-Pérez, who explained to me that despite the real possibility of spotting a jaguar there, the problem with both national parks was connectivity. One of Panthera's primary goals in Costa Rica was to safeguard jaguar travel routes, connecting Santa Rosa to similar areas north across the border in Nicaragua, and Tortuguero to the heart of Costa Rica's rugged Talamanca Range, in order to avoid the genetic dead ends that had kept Rabinowitz up at night.

Consequently, one of Panthera's goals was to connect Corcovado, via a wild green corridor, stretching east to the rugged and remote Talamanca Mountains and the million-acre La Amistad International "Friendship" Park. Extending from northern Panama into central Costa Rica, La Amistad is part of what Panthera calls its critical "backbone corridor." The small, privately owned *fincas*, which I visited, just outside Corcovado National Park, were key to that connectivity. Ordinary farming families had embraced conservation goals and were prospering from ecotourism. But at the eastern end of the forest reserve, tourism had been sluggish, and progress among landowners to adopt conservation measures had been achingly slow. A jaguar hoping to travel from Corcovado to the Talamancas would have to negotiate its way through the reserve and then across the 34,600-acre Piedras Blancas National Park, confronting an assortment of obstacles, from jaguar-poaching farmers to roads and illegal logging.

After breakfast, Dani, Jason, and I waited for Champeon, who as soon as he stepped off the porch was already bound for the forest. Without missing a step, he slipped his machete from its scabbard and waded deftly into the jungle, slashing vines as he walked. Sure-footedly, he set the pace, a moderate, energy-saving one that he likely could keep up for the entire day. As we walked, he pointed out damselflies, a tiny poisonous dart frog, philodendra, and enormous avocado and fig trees, battling for sunlight and choked by hanging vines.

As we ascended a small rise, an overpowering smell filled the jungle. Dani explained that the scent wall we had just walked through was tapir pee, which excited everyone. The tapir, a comical-looking creature with a floppy, elephant-like snout, was considered an indicator species, a sure sign of the ecosystem's health. Champeon halted for a moment to show us the three-toed imprint of the bulky tapir's back foot embedded deep in the mud.

At the top of the hill, we stopped for our first water break next to a giant ficus tree with enormous, balancing roots that prevented it from toppling over. In the shallow soil, the roots stretched out laterally in every direction. After an hour of walking, I was drenched in sweat—Champeon, Dani, and Jason seemed impervious to the heat—but my spirits were high. That's when a band of angry spider monkeys, eager to express their displeasure at our intrusion, came crashing through the trees and descended from the canopy. I looked up just as one hurled something in our direction.

"Don't look up," Dani warned, a bit too late. Sticks and seed pods rained down on us, as the monkeys continued to scream and vent.

After escaping the monkeys, our route took us up a steep muddy hill, made even muddier by the recent rain. I slipped and slid and lost my balance far more than the others, who after two weeks in the hinterlands of Corcovado had already acquired their "jungle legs." At one point, I fell to my knees and made the mistake of reaching out for a handhold. I knew instantly it was the wrong reaction. Whatever I'd grabbed was enveloped in thorns, spikes, and tiny red fire ants.

I yelped, and when Dani crested the hill, he looked back. By that time, I had scrambled to my feet again. When I reached Dani, he consoled me saying that in Corcovado they had a motto: "You have to fall to be a ranger."

By the time we reached the first camera trap, my hand was throbbing. Jason could tell that I was in pain, and he grimaced as a gesture of empathy. While Champeon knelt down next to the tree to which the camera trap was attached, I fished through my backpack, remembering that I had hydrocortisone cream in my first aid kit. When I spread a dollop of cream over my palm and fingers, the relief came quickly.

Champeon handed Dani the SD memory card from the camera. After inserting another card and locking the box, Champeon got down on all fours and crawled into the camera's field of vision. A red light blinked, assuring him that the infrared sensor was working. Then Champeon took out a small notebook and took careful field notes.

When he stood and shouldered his pack, he said, "*Vamonos*." But by the tone of his voice, I could tell that he was asking me if I was up to it. I nodded and we were off again, navigating a strenuous route that wandered up and down steep hills and deep ravines cluttered with rocks, fallen trees, half-buried in mud and clay, and thick, hip-high ferns. Whatever trail Champeon was following was indistinguishable to me. Not once did I see him consult a map, a compass, or a GPS. And when he walked, he didn't meander; he moved unerringly and with purpose.

Once we reached the second camera trap, near an enormous fig tree with dozens of strangler trunks that looked like giant anacondas, I was winded and so bathed in sweat that I took off my shirt and wrung it out.

After the second camera, we checked a handful of others and then headed back to the station, where I spent the afternoon tending to my hand and small wounds where thorns had punctured my skin. But my biggest problem was ticks. I'd apparently stumbled into a nest. Despite wearing gaiters and tucking my pants into my socks as a precaution, somehow dozens of ticks had found their way inside and up my legs. From my knees up, I was covered in brown spots. Fortunately, most of the ticks hadn't had the time to embed themselves, and with a sharp, heated knife blade I was able to scrape them off.

The following days unfolded in much the same way. I followed Champeon and his team through the jungle, never quite capable of getting my bearings but trusting always in Champeon's sense of direction. We checked cameras and returned in the afternoon, whereupon Dani and Champeon spent the remainder of the day in Champeon's office on the second floor of the main station building, scanning photos from the cards they'd collected.

One afternoon I was sitting in the dining area reading, when Dani summoned me. I followed him up the stairs, noting that each tread had the imprint, or pugmark, of a jaguar's paw. The imprints led to Champeon's little office on the second floor, where he was sitting in front of his computer. A figurine of a jaguar sat on the corner of his desk.

I walked in, and Champeon was smiling. I heard him click his mouse. When the photo came up, Dani whooped, and Champeon slapped his hand gleefully on his desk.

"*Que lindo,*" Champeon said, marveling at the photo.

It was a picture of a powerfully built jaguar with the low-slung belly of an alpha male and a head as big as a basketball. I could feel the electricity rip through the room. This was the moment that jaguar researchers wait for. Short of seeing one in the wild, this was as good as it got—setting the camera trap in exactly the right spot and being rewarded with a candid close-up of one of the world's most obscure predators. The photo was so clear and clean that the jaguar's coat looked as if it were shimmering.

"Macho Uno?" I asked Dani.

He shook his head. "A new one. It's the first time we document this one."

Surely, that was also part of the joy—knowing that Corcovado was producing new jaguars and knowing that the photo was taken in the light of day. If hunted, predators and prey alike become almost exclusively nocturnal.

Champeon continued scrolling. He stopped briefly at a few photos of pumas, but the jaguar was clearly the big prize. The next time he stopped, I saw a large, tightly packed herd of white-lipped peccaries. A few minutes later he halted at an image of a tapir and pointed at a large scar across its midsection.

"*Tigre?*" I asked.

Champeon nodded and traced the wound on the screen with his finger. The scar was a reminder of the kind of violence that happens in a place like Corcovado, where a jaguar, if its timing is perfect, can kill a three-hundred-pound tapir in seconds and has the brute strength to drag it through the jungle or, as is sometimes the case, fails in its attempt and leaves the tapir with deep scars.

Champeon again scrolled quickly, his eyes registering the details well before mine did. For me, the photos were just a blur of jungle.

Then suddenly he stopped. "*Ocelote*," he said.

It was a beautiful animal that looked like a miniature jaguar.

I spent another forty-five minutes in Champeon's office and was treated to photos of another puma, an oncilla, and a jaguarundi. Dani proudly pointed out that we'd seen five species of cats, a remarkable achievement for any environment and any camera trap program.

But Champeon had saved the best for last. I would have accused him of staging the event for the sake of drama had I not known that he was looking at the images for the first time himself. He was scrolling fast and then again stopped abruptly. Something had caught his eye. He clicked on it, and there it was.

Champeon nearly jumped out of his seat. He turned and gave Dani an energetic fist bump. A pregnant female jaguar.

For Champeon, it was like striking the Madrigal River mother lode.

CHAPTER 13

A Dream Undone?

ACCORDING TO HIS FRIEND JANE ALEXANDER, RABINOwitz went to Thailand, intending to save the Huai Kha Khaeng Wildlife Sanctuary from "illegal hunting, tribal people, and the rich and powerful Chinese."*[1] Instead, he encountered an illegal wildlife trade so vigorous, it left him bereft of hope. He was "going to sleep at night to the sounds of gunshots," and during the day, he saw "countless body parts of animals in the marketplaces."[2] In fact, the only clouded leopards he saw in Thailand were those that villagers had trapped and whose skins they tried to sell him.[3]

Struggling to understand what motivated the poachers, he participated in what he called "opium council(s)," where the opium pipe was passed around the circle.[4] Ultimately, though, he realized that the wildlife trade was so deeply entrenched, and the government's lack of political will was so fixed, that he was powerless to save Thailand's tigers or its clouded leopards. Jane Alexander wrote, "The early enthusiasms with which he had greeted his [Thailand] assignment had eroded." For Rabinowitz the magic "was . . . long gone."[5]

---

* Some Chinese companies openly advertised their ability to supply animals listed in Appendix I and II of CITES.

Despite his inability to achieve his conservation goals in Thailand, on a personal level the country had been good to him. There he met his wife to be, a beautiful and talented Thai woman named Salisa Sathapanawath, who was studying at Mahidol University in Bangkok for her master's degree.

Salisa was cut from the same adventurous cloth as Rabinowitz. She was an independent-minded scientist who reveled in the wild, spending months in the field conducting surveys on the ecological implications of government infrastructure projects, and on the status of various diseases, especially malaria and dengue fever. She too had penned scientific articles and had written a book on the wild cats of Thailand.* So months after meeting, when Rabinowitz proudly presented her with his book *Jaguar* and told her, in a moment of immodesty, that he was becoming something of a "famous" scientist, she was not impressed. In fact, after reading a draft of the book, she told him that she didn't think he was a particularly gifted writer. She admitted, too, that she wasn't fond of its frequent personal details, and that in her opinion he had stepped over the line between science and self-revelation.

Ultimately, the *Los Angeles Times* would call the book "immensely readable," and Tom Miller of *The Washington Post* wrote that in *Jaguar* "the tension between man and beast becomes startlingly vivid."[6] Rabinowitz still had much to learn about being a writer—Miller wrote that Rabinowitz's prose occasionally bordered on the "hyperbolic" and was rarely "graceful"—but at the end of his review, he penned the sentence that might have singlehandedly launched Rabinowitz's long writing career. "Fortunately," he wrote, "his entries didn't just end up in scholarly periodicals or letters home, and we are the better for it."

Inside the Wildlife Conservation Society and the world of jaguar biology, however, Rabinowitz's book met with mixed opinion. Later, he wrote that colleagues were "clearly embarrassed" for him.[7] They

---

* Salisa, who flourished in the field, also accompanied Alan and George Schaller on trips to Laos, northern Vietnam, Myanmar, and northern Thailand in the late 1990s.

wondered why he hadn't written a scholarly tome more like that of his mentor, George Schaller. They wanted him to stick to the science and dispense with the romance, the failed love affairs, the snake attacks, the death-defying adventures, and the self-recrimination. Schaller's journals could be candid, but his books steered clear of the personal. Rabinowitz, on the other hand, had deliberately portrayed himself as something other than an "objective scientific observer." He had written a kind of bildungsroman. He didn't know it at the time, and the criticism surely stung, but he was carving out his own literary brand, one indebted to Schaller, but very much his own.*

Despite her criticism of his book, Rabinowitz and Salisa grew fond of each other, and over a year later, on a trip to the Salawin Wildlife Sanctuary in the far northwest of the country, where Rabinowitz was tracking tigers and Salisa was doing her own survey for the United Nations Development Program and the World Bank, Rabinowitz proposed, kneeling and presenting her with a brass ring that he'd bought from a street vendor in Chang Mai. Salisa accepted, conditionally, explaining to Rabinowitz that she had been admitted to a Ph.D. program in the anthropology department in genetics at Columbia University and wanted to finish her degree before they married.

In the summer of 1992, when Salisa was just halfway through her Ph.D. program, the two married on Jane Alexander's sprawling lawn in Putnam County, New York. The only guests were Alexander and her husband and Rabinowitz's longtime friend Howard Quigley and his wife. All agreed that Rabinowitz had, indeed, found a wonderful partner, someone as sharp and curious as he, and also a woman who had the temperament to deal "with a man of so many moods."[8]

---

* While in Belize, Jane Alexander read Rabinowitz's journals and became convinced he had the "makings of a book." She introduced him to her literary agent, and ultimately, he signed a publishing contract with Island Press. In an updated version of *Jaguar*, he acknowledged that some of his friends and colleagues who read the book were embarrassed for him.

NOT LONG AFTER HIS MARRIAGE TO SALISA, and still dispirited by his experience in Thailand, Rabinowitz set his sights on a country that had long fascinated him, Myanmar.

Rabinowitz had already established a reputation as a biologist willing to take on the toughest projects. He had spent the decade after Belize in far-flung parts of Southeast Asia, studying tigers and clouded leopards in Thailand and trudging through the interior of Borneo in search of scarce Sumatran rhinos. He established Taiwan's largest protected area and its last piece of intact lowland forest while in search of the Formosan clouded leopard, and he executed the first biological survey of the obscure Annamite Mountains between Laos and Vietnam.

Myanmar measured 261,228 square miles, almost thirty times the size of the diminutive Belize. It was also one of the ten poorest countries in the world. But it did possess a staggering biological wealth, from its dense rainforests in the south to the rugged, icebound Himalayas—in Sanskrit, literally, "abode of snow"—far to the north. However, only 1 percent of the country's land was protected. Worse yet, the new government's economic policies encouraged the exploitation of natural resources.

But upon arriving in Myanmar, Rabinowitz experienced an "intense feeling of 'rightness' that crept up from deep inside [his] gut." He had grown deeply cynical about the state of Asia's wildlife, but he hoped that Myanmar could convince him that he "still cared enough to try to make a difference in the world."[9]

Some conservationists denounced him for his desire to work in Myanmar. Its generals had savagely quelled a pro-democracy uprising, killing between 3,000 and 10,000 protesters, and imprisoned hundreds more. But Rabinowitz was quick to justify his decision, explaining—like Schaller—that conservation superseded politics. "If wildlife conservation has first to be considered through political filters, then where should I work in the world?" he asked rhetorically, adding, "And when did animals get to vote and decide what governments they must live and die under."[10]

As a gesture of support, Schaller joined him on his first in-country expedition. The two friends traveled fifteen hundred miles north of Yangon to the 232-square-mile Htamanthi Wildlife Sanctuary, which no Westerner had laid eyes on in decades. What they discovered was a beautiful landscape nearly devoid of animals. The hunters admitted they no longer saw rhinos and rarely came across tigers. The "empty forest syndrome" that Rabinowitz had encountered in Thailand, where despite being "deep inside one of the biggest protected forests in this part of the world . . . [he] couldn't even hear the birds singing," had penetrated the most remote parts of Asia.[11] Htamanthi was a paper sanctuary, a classic case of a protected area that had never been properly defended.*

Upon returning from Htamanthi, Rabinowitz requested government permission to mount a months-long biological reconnaissance expedition in the vast and untamed region around Mount Hkakabo Razi. Far to the north, some still regarded it as the territory of the fearsome yeti, and no American had entered in half a century. While waiting for approval, he trained like a man possessed in his makeshift gym for the longest and most arduous trek he'd ever attempted, dedicating himself to a regime of boxing, martial arts, weightlifting, and running. Surrounded by inspirational posters, the words of Rudyard Kipling echoed in his head: *This is Burma, and it will be quite unlike any land you know about.*[12]

As Rabinowitz exercised to the point of exhaustion, one thing continued to gnaw at his conscience. He couldn't escape the feeling that he had sold out the Maya, who had been evicted from their traditional lands in the Cockscomb Basin. But in early 1997, while waiting to hear about the expedition to Mount Hkakabo Razi and writing the foreword to a book about the Cockscomb, he stumbled upon the words of his Maya friend Ignacio Pop, Cockscomb's first warden. "I see how my children will never see these wildlife except we save it,"

---

* Jane Alexander attributed the disappearance of animals to a "culture of killing" that "was growing and spreading like a cancer across Asia." Alexander, *Wild Things, Wild Places*, 48.

Pop wrote. "I always thank for what Alan has done. Without him, it would all be gone."[13]

Rabinowitz would ultimately make two daring and dangerous expeditions to northern Myanmar, one of which he called the "most physically exhausting journey of [his] life."[14] *The New York Times* took notice, running an effusive article titled "Alan Rabinowitz; Indiana Jones Meets His Match in Burma."[15]

The article began,

> In 16 years with the Wildlife Conservation Society, Dr. Rabinowitz has tracked and captured jaguars in Belize, tigers in Thailand and leopards in Borneo and Taiwan. He has lived with Mayan Indians in Central America and jungle tribes in Southeast Asia.
>
> Most recently he trekked for a month to reach one of the world's most isolated areas, in northern Myanmar (formerly Burma), crossing mountain torrents on hastily built bamboo bridges that would have given the unflappable Dr. Jones pause.
>
> Moreover, the treasures that Dr. Rabinowitz has helped save, among them a 1,472-square-mile national park in northernmost Myanmar and Taiwan's largest nature preserve, are infinitely more valuable in the eyes of many than possessions like the fictional Lost Ark.[16]

Based on his expeditions, Rabinowitz feared for the future of the tiger in Myanmar. He considered the country's tiger population to be declining precipitously.* Even in the Hukawng Valley, he suspected that hunters, equipped with tremendous crossbows and poison arrows, were killing tigers faster than they could reproduce. The hunters sold their parts to Chinese traders who often paid them in

---

* As he had done in Thailand, he again chose to use tigers as his "proxy for a relatively healthy environment" because they required "large areas and an abundant prey base." Rabinowitz, "New Strategy for Saving Big Cats."

a curious currency—with salt—which among the sodium-deprived people of northern Myanmar was more valuable even than cold, hard cash.

Months after submitting his proposal for the region, Rabinowitz was stunned when the director general of Myanmar's forest department, alarmed by his report on the state of Myanmar's tigers, asked him to fly to Yangon to discuss expanding the sanctuary to include the entire nine-thousand-square-mile Hukawng Valley. Overwhelmed by the request, he boarded the first plane to Myanmar. He had built his career on championing big, bold projects, but this seemed nearly "incomprehensible" to him.[17]

After a series of successful meetings, he returned home to Mahopac, New York, elated but exhausted. The years of strenuous treks and tiring diplomacy had taken a toll on his health. Jane Alexander described him as one of the strongest men she'd ever known, but she could tell that her friend was suffering physically and emotionally.

Just weeks after arriving back in the United States, he was plagued by night sweats, achiness, and a general malaise. At Salisa's urging, he made an appointment with his doctor who assumed that he'd likely contracted a virus overseas. Viruses and parasites were hazards of the profession, and Rabinowitz had always taken them in stride. But a few weeks after his first appointment, he returned, concerned that the symptoms, far from disappearing, had grown worse. His doctor suggested a routine blood test.

Days later Rabinowitz again met with his physician to discuss the results—a significant elevation of his white blood cell count—and to have a bone marrow sample taken that very day. Later, at home, Salisa's eyes filled with tears as Rabinowitz explained the details of the doctor's diagnosis.

When he returned a week later, Salisa accompanied him. Despite the preliminary diagnosis—chronic lymphatic leukemia, a cancer for which there was no known cure—his doctor explained that his cancer was incurable but often slow-moving.

Despite his doctor's optimism, over the next weeks, Rabinowitz fell into a "deep, pitch-black hole of sadness."[18]

CHAPTER 14

A Bold Conservation Model

AS HE WAS STRUGGLING WITH HIS CANCER DIAGNOSIS, Rabinowitz became increasingly fearful about the state of the jaguar and jaguar conservation. He had assumed that the work he'd done in Belize would spur more groundbreaking research. But that wasn't the case. In fact, most jaguar biologists were still relying on an outdated distribution map published in 1987. In the intervening years, precious little new information about the big cat had emerged.

Despite his cancer, he rededicated himself to protecting jaguars across the Americas. Their territory was shrinking year by year as the human population of Central and South America boomed. What's more, across what he would come to refer to as the jaguar's "cultural corridor," where the big cat had been mythologized for thousands of years, jaguars were often reviled by the people among whom they lived.

In March 1999, during a break from his work in Myanmar, Rabinowitz had attended a conference in Cocoyoc, Morelos, Mexico, called "Jaguars in the New Millennium" organized by the Wildlife Conservation Society and the Universidad Nacional Autónoma de México and funded, in part, by his friends Liz Claiborne and Art

Ortenberg.* Thirty-five specialists gathered there for the first range-wide meeting of jaguar biologists.

Based on an enormous amount of data, which never before had been assembled for the jaguar, the group established that the big cat was still present in 46 percent of its historic range. However, in 20 percent of that range actual numbers remained a mystery. Using the data, the group zeroed in on fifty-one havens, consisting of stable big cat populations and an abundance of viable habitat, which included reliable prey populations. Alarmingly, they discovered that although 67 percent of those havens enjoyed some form of protection, only 3 percent were sufficiently protected.

Almost a full year after the gathering, Rabinowitz finally found the time to read a rough draft of the book—*El jaguar en el nuevo milenio* (The Jaguar in the New Millenium)—that the conference produced (which included thirteen papers from Brazilian biologists mentored by Peter Crawshaw). One in particular, titled "Evolution and Genetics of Jaguar Populations: Implications for Future Conservation Efforts," would prove a revelation.[1]

Great Britain's most prominent mammologist, Reginald Pocock, had divided jaguars into subspecies in 1939, and those divisions had held for over half a century. In fact, well into the 1990s, biologists believed that subspecies of jaguars existed, that genetically the jaguar in Argentina was different from one in Arizona. In his laboratory in the basement of the Natural History Museum in London, the pioneering Pocock studied fifty-nine jaguar skulls, noting their similarities and differences. Biologists hadn't even considered contesting his conclusions: that eight distinct subspecies of jaguar existed.

It wasn't until 1997 that the biologist Shawn Larson published

---

* Jaguar Canada (the company that makes the luxury cars) made a large charitable donation that helped to create the Cockscomb Basin Wildlife Sanctuary. It also helped to underwrite the 1999 Mexico meeting. Highlighting the company's financial contribution, Rabinowitz said, "Without this type of support, we're not going to be able to save these animals." In 2003 the company created the Jaguar Conservation Trust, which helped to fund programs and projects to save wild jaguars.

a bold article, based on her master's work on jaguars, in an esoteric journal called *Zoo Biology*, that anyone dared to call Pocock's long-standing divisions into question. Larson realized that some of Pocock's subspecies designations were based on a single skull, so she examined data from 170 jaguar skulls from U.S. museums. Though she found superficial differences in populations, she discovered "no significant taxonomic differences between jaguars throughout their range."[2] The greatest variation—though not statistically significant— occurred between jaguars from northwestern Mexico and jaguars from the Amazon.

Larsen went on to suggest that jaguar populations should be managed range-wide, as a single species without subspecies. Rabinowitz rejected her findings until he read a paper by Brazilian Eduardo Eizirik (and his collaborators), "Evolution and Genetics of Jaguar Populations," that reminded him of Larson's original article. A subsequent paper by Eizirik emphatically reinforced Larson's proposition. It urged conservationists to develop strategies for jaguar management that were aimed at maintaining "high levels of gene flow over broad geographic areas."[3]

When the Laboratory of Genomic Diversity in Frederick, Maryland, part of the National Institutes of Health, analyzed jaguar mitochondrial DNA from blood and tissue samples collected throughout the Americas, researchers determined something remarkable: No jaguar group had split off into a true subspecies. They saw clear evidence of a continued gene flow, establishing the jaguar as the only large, wide-ranging carnivore with no subspecies. Rabinowitz was convinced that someone had made a "mistake" and stubbornly resisted the idea.[4] But eventually, he embraced the irrefutable: From Sonora Mexico to the Iberá of northern Argentina, jaguars had been wandering great distances to breed with each other. Capturing his mood at the moment of his realization, Rabinowitz would later exclaim in *An Indomitable Beast*, "There was only one jaguar!"[5]

The discovery meant the jaguar had miraculously navigated— and was continuing to navigate—its way across mountains, rivers, deserts, and in the modern world, a hostile landscape fragmented by roads, canals, and endless agricultural fields, avoiding "the black abyss of humanity." Each jaguar had its own markings, as unique as a human's fingerprint, but across thousands and thousands of miles, they all possessed the same DNA.

With this new and momentous information, Rabinowitz was forced to reconsider his entire strategy: "Saving a wide-ranging mammal species throughout its entire range [had] never been attempted before."[6] The jaguar's range had to be managed as a "single ecological unit."[7]

Given this new paradigm, the critical challenge was figuring out how to connect the viable jaguar populations in northern Mexico with those in Panama and Bolivia and southern Brazil. To accomplish this, he would need to follow the jaguar's lead and protect the routes it was already using.*

---

* Rabinowitz had two models to guide him. One was his WCS colleague Archie Carr's program, Paseo Panthera, or "The Path of the Panther," to create a corridor from the Darién Gap to the Selva Maya of Guatemala, Belize, and Mexico. The other was the Mesoamerican Biological Corridor.

CHAPTER 15

Green Gold

NEAR DAWN I HEARD A RAP ON THE SIDE OF THE FOUR-by-four, the high-pitched sound of Nick McPhee's small flashlight on the metal of the vehicle. Then I heard McPhee's voice. "Jaguar tracks," he said excitedly. Before José, our driver, could stop, McPhee was on the ground. From up above, I gazed at the road, but I didn't see the tracks at the road's edge until I climbed down from the roof of the Toyota Hilux. Then I saw a series of perfectly formed imprints, etched deep into the mud.

"A big male," McPhee said, kneeling and pointing out its round shape and the spread of the toes. When Ezra, my friend, part-time research assistant, fluent Spanish speaker, and ecologist, knelt next to McPhee, I heard the speedy *tick tick tick* of his camera. We hadn't spotted a jaguar, despite all our searching, but a fresh track like this was almost as good. Besides, I had come to understand that catching a glimpse of a jaguar was a case of pure and simple serendipity.

"That cat just passed through here," McPhee said rapidly, as if trying to catch his breath, a wad of coca leaves bulging from his cheek. "He could be just ahead of us." Tracking now, McPhee walked quickly, athletically. As a young man in Australia, where he was raised, he'd joined the army and served in Iraq for six months, and he later worked in Afghanistan as a contractor. He played a lot of rugby and still looked and moved like a rugged, broad-shouldered athlete, with a nose as flat as that of a bare-knuckle brawler.

We followed the fresh tracks on foot and by vehicle until the jaguar entered a marsh and its tracks were no longer visible. The marsh extended far to the south and was bordered on its eastern end by flat agricultural fields that stretched seemingly as far as the sky. Worship of the soybean had overtaken much of eastern Bolivia. In Santa Cruz, the largest of the country's nine departments, nearly half the size of the state of Montana, the tragic destruction of jaguar habitat was occurring at an alarming rate.

When I arrived in Bolivia, I hadn't known what to expect. Apart from the reading I'd done before my trip—and having watched *Butch Cassidy and the Sundance Kid* as a boy—I knew very little about the country. Wildlife biologists consider Bolivia to be a big cat stronghold but an increasingly fragile one because of a growing trade in jaguar parts, uncontained fires, aggressive agribusiness, Chinese investment, a government bent on a course of development that was at odds with the environment, and the inexorable loss of connectivity on which jaguars depend.*

Although the dogged McPhee was still intent on finding a jaguar, my visit to Bolivia was not about seeing a big cat. I had come here, in part, because Alan Rabinowitz hadn't. Although he and his team had traversed (ground-truthed) portions of the Jaguar Corridor, and he later visited parts of the corridor on what he called the Journey of the Jaguar, he'd never been able to make it to Bolivia. But he was both fascinated with the country and deeply disturbed by it, especially by reports of both the killing of jaguars for domestic markets and for a more sophisticated, pernicious, and abhorrent form of international trafficking. He had witnessed both phenomena during his work in Southeast Asia. In *An Indomitable Beast*, he wrote, "I was confronted with something I was not prepared for—Asian medical

---

* A forceful case preserving the dwindling corridors linking Bolivia to the Brazilian Pantanal and the Amazon may be found in Boron, "Jaguar Densities."

practices . . . that highly valued the parts of almost any animals that flew, walked, or crawled the earth, especially if it was big and strong. Tigers topped the list, with everything from tiger blood to eyeballs to dried feces being used to promote vitality or to cure a wide variety of ailments."[1] After he transitioned from tigers back to jaguars, he came to fear the emergence of an international trade in jaguar parts that, as tigers disappeared from the wild, was accelerating across the Jaguar Corridor, especially in Bolivia.

To learn about the state of the jaguar in Bolivia, I had sketched out an itinerary that would take Ezra and me to Jaguarland, where McPhee was attempting to build an ecotourism business around the big cat; to the San Miguelito Jaguar Conservation Ranch, where Duston Larsen and Anai Holzmann were transforming a traditional cattle operation into a multipurpose ranch with jaguars as the main objective; and to the sere and isolated Kaa-Iya del Gran Chaco National Park, a jaguar safe haven in the country's far south.

Prior to seeing the jaguar prints so perfectly embedded in the sand, McPhee, José, Ezra, and I had spent much of the previous two nights driving the rutted back roads, scanning with our spotlights the deep ditches, ponds, and tangled edges of the forest for jaguar eyes. We'd been jostled and nearly thrown from the truck's roof a number of times, but we'd seen jaguarundis and sleek ocelots. McPhee was frustrated, however, because documenting jaguars was essential to his plan for Jaguarland, an archipelago of remnant forests surrounded by expanding fields of soy. By identifying the existence of multiple jaguars, McPhee hoped to be able to convince the landowners that the forests were worth preserving.

But the weather had conspired against us. First, the rain blew in, and then a rare cold front with persistent winds that likely kept the big cats bedded down. The trip was anything but a disappointment. Ezra, who had begun birding in his native Evanston, Illinois, as a boy was thrilled to add seventy birds to his life list. We'd even seen a prehistoric-looking giant anteater lumbering muscularly through the grass at the side of the road, a sight almost as rare as the sight of a jaguar.

The anteater, an improbably beautiful animal, resembled an overgrown wolverine, with its bushy tail and large forefeet, or perhaps a black bear, which explains why they were once referred to as ant bears. The anteater was close enough for us to see its long, curved snout, which it uses like a suction hose, and the bold white racing stripes running along its torso. What we couldn't see were the rapier-like claws that it used with great efficiency to dig out anthills and to protect itself from predators. McPhee explained that even jaguars feared anteaters. Though they have no teeth, they are capable of delivering violent, lightning-quick blows that can lay open a big cat in seconds. McPhee said he'd seen a few jaguars bearing ugly scars, which were likely the result of an unsuccessful attack on an anteater.

Despite the thrill of spotting the small cats, the anteater, and an astounding display of avian biodiversity, McPhee was unhappy with the trip. Back in the city of Santa Cruz de la Sierra, or simply Santa Cruz, five hours to the south, his friends joked about his obsession with jaguars. He had joined forces with Conservación Loros Bolivia, an organization dedicated to saving the country's endangered species, and in his gruff, rapid-fire Aussie accent, he could wax enthusiastic about hyacinth and blue and green macaws, looking like colorfully painted, long-tailed kites, stately jabiru storks, and the predatory skills of tayras. But the adrenaline surge he got from spotting a jaguar was something to behold. Although he'd seen big cats more than one hundred times, sometimes when he laid his eyes on one, he still forgot to breathe.

The first time McPhee ever saw a jaguar outside an Australian zoo, it was 2012, and he was in Bolivia's Madidi National Park, one of the most biologically diverse places in the world, in the upper Amazon River basin. That sighting was only a glimpse, five seconds at the most, before the big cat drifted from the sun-splashed riverbank into the dark jungle. But it was all he needed. From that moment on, he was hooked and decided to devote his life to learning as much about jaguars as he could.

After we lost the cat tracks, we turned north onto a muddy road, passing one makeshift settler encampment after another. Here the problem wasn't big agriculture, but rather the arrival of poverty-stricken people, desperate for land. Most of them had been here only long enough to erect rough shelters covered by bright plastic tarps. McPhee said they came in groups, a dozen or so men with axes, shovels, and picks and sometimes heavy equipment. Their method was an unmerciful one. First they slashed the land. Then they burned it and slashed it again until the vibrant forest looked frail and lifeless. Once the land was sufficiently cleared, they planted crops on the poor jungle soil. They'd get a harvest for a year or two and then move on.

McPhee explained that the newcomers were landless Altiplano settlers, Indigenous highlanders from the Andes in the west, drawn to the Santa Cruz department by Evo Morales's promise of free land in the east.

Like many people in the Santa Cruz department of Bolivia, McPhee disliked Morales fervently. Morales was no longer the country's president, having fled to Mexico in 2019 for political asylum in the wake of what he described as a "coup" by the country's military and other forces opposing his presidency. But McPhee believed that his ghost, and his policies, still haunted Bolivia.

He was particularly fond of a scathing article in *The Guardian* that, quoting the government's detractors, claimed Morales was a "Murderer of Nature" for legislation he had passed in July 2019, encouraging peasant farmers to employ slash-and-burn methods to create pasture and arable land.[2] According to McPhee, acres upon acres of smoking and denuded rainforest was the inevitable outcome of that policy. Contrary to its impressive image on the international stage as a protector of biodiversity and marginalized social groups, the Morales government, McPhee claimed, had quietly approved the rapid expansion of agribusiness, as well as a host of mining and road development projects, while simultaneously undermining conservation efforts in some of the country's most sensitive protected areas and Indigenous Territories.

For two years, fires started by settlers had raged across the country, most of them blazing through what some considered the politically exiled eastern part of the country, rather than Morales's beloved

Andes. Morales's critics said that for weeks he ignored the rampaging fires, refusing to accept either international aid or help from Argentine firefighters. By the time the fires finally subsided, more than 4 million acres, a whopping 12 percent of the country's Chiquitano forest, the world's best-preserved tropical dry forest, had been obliterated. According to Alfredo Romero-Muñoz, a Bolivian doctoral researcher at Germany's Humboldt University, and his associates, Bolivia had plans to "quadruple agricultural land into forested areas by 2025." He added, "If these trends continue, Bolivia could lose almost all of its remaining forests by 2050, together with immeasurable biodiversity loss, including emblematic animals like jaguars."[3]

As unnerving as it was to see once-thriving forests slashed and burned, what we saw on the way back to Santa Cruz made what the settlers were doing look inconsequential by comparison. The Santa Cruz department accounted for 70 percent of the country's agricultural output. Large companies had transformed the landscape into an agribusiness utopia, carving it into perfect geometries, without even shelter belts to stop the wind and hold the soil. The land was largely flat, but when there was contour, in the coulees and draws, forests remained, or at least remnants of what were once forests.

Few countries in the tropics have seen their forests fall as fast as Bolivia has. Bolivia is one of the most ecologically and biologically diverse countries in the world, and almost half its national territory (131 million acres) is covered by forest, but that forest, so vital to maintaining its diverse ecosystems and healthy populations of jaguars, is disappearing fast.*

SOY IS REFERRED TO AS "GREEN GOLD" in Bolivia. The following morning, en route to the San Miguelito Jaguar Conservation

---

* In addition to containing large untapped deposits of oil and natural gas and veins of gold and uranium, the Santa Cruz department accounts for 70 percent of the country's agricultural output and a third of Bolivia's GDP. Becoming the nation's breadbasket has not come without consequences.

Ranch with Larsen, Holzmann, and Damián Rumiz, a prominent wildlife researcher at the esteemed Noel Kempff Natural History Museum, we saw endless fields of soy, separated by frail shelterbelts of trees. A new law stipulated that the shelterbelts had to be a minimum of fifty meters wide, but the law was rarely enforced.

Two and a half hours into our trip, we turned off the paved two-lane highway that ran east all the way to Corumbá, Brazil, onto a dusty road. The day's heat rose slowly from the fields. We hadn't gone far when we began to see horse-drawn buggies, driven by men wearing dark bib-and-brace overalls, blue shirts, and straw Panama hats. Others, looking like something out of a children's fairy tale, were pushing large scooters, outfitted with bicycle wheels and baskets.

"Old Colony Mennonites," Duston Larsen said.

The Old Colonies, Larsen explained, reject most of the trappings of the modern world. Farmers still milk their cows by hand from wooden stools. Though they live insular, largely traditional lives, Bolivia's nearly 100,000 Mennonites, who came to the country fifty years ago from Mexico, are a force to be reckoned with in the agricultural world. They are pivotal producers in a robust regional economy centered on soy. In 2015, with seed varieties adapted to the tropics, Bolivia produced over 2 million metric tons of soybeans, valued at over $1 billion. By the sweat of their brow and biblical exhortations to transform wilderness into civilization, Mennonites contributed to over a third of that production. What's more, with soy production in Brazil, Argentina, and Paraguay reaching critical mass, Bolivia intends to triple its lands devoted to agriculture by 2025. Bolivia's Mennonite farmers will surely play a key role in that expansion.

Rumiz explained that though some of the Mennonite farmers had broken ranks and were slowly embracing conservation objectives, most persisted in large-scale soy production, which produced fields that—without the wetlands, forest, and riparian galleries—were dead zones for jaguars. Larsen lamented that those dead zones had already reached the gates of San Miguelito, whose seven thousand acres played the role of the proverbial rescue ark in an area surrounded by a dystopian landscape of agricultural fields, stretching for dozens of miles in every direction. San Miguelito was, indeed, a refuge. Despite Larsen's entreaties, his brothers, who had taken over their portions of the ranch, had leveled their forests and sold out to soy.

Holzmann—a biochemist, jaguar advocate, and Panthera consultant—said that San Miguelito had signed what she called "a peace treaty" with predators. "Our goal," she added, "is to show ranchers that jaguars are worth more alive than dead." She explained that as part of that strategy, they had signed agreements with the provincial and local governments to create La Ruta del Jaguar (The Jaguar's Path), an "eco-ethno-tourist" effort that used the jaguar as a symbol to promote income opportunities for local communities in the area. The project, she said, focused on educating and informing people about how to prevent human-carnivore conflict and jaguar-cattle predation. At the same time, it would use the stirring image of the jaguar, and its very real presence on the land, to create income from wildlife tourism that could compensate Indigenous farmers for cattle losses. La Ruta del Jaguar gives local communities a stake in protecting jaguars and taps into the historic admiration for the big cat.

As the truck kicked up clouds of dust, and Larsen swerved to avoid potholes, he picked up on Holzmann's narrative. "Farmers and ranchers shoot big cats because of bad cattle management. We lost fifty-two animals in 2013, and in 2018 we lost four. In 2020 we lost only two calves and a horse. What are we doing differently?" he asked rhetorically in his booming voice. "We're using Panthera anti-predation strategies, just simple steps that any rancher can take to protect his animals from predators."

Those anti-predation methods, which have been explored and refined on Panthera's model ranches in Brazil, Costa Rica, and Colombia, begin with protecting the animals that jaguars and pumas like best: defenseless calves. Emulating tried and tested techniques, Larsen and Holzmann corral their calves and give them supplemental feed when their lactating mothers are sent out to pasture in areas frequented by big cats. They have also introduced enormous water buffalo and a feisty horned cattle breed known as Criollo to help protect the herd.

As Holzmann and Larsen described the San Miguelito model, and especially the involvement of the local Indigenous population, I kept thinking about how enthusiastically Alan Rabinowitz would have reacted to their description of their jaguar-friendly enterprise. It was exactly the message he had been preaching across the jaguar range.

As we neared the ranch, Larsen gave us a brief San Miguelito history lesson. In 1968 his father, Ronald Larsen, left eastern Montana to pursue a scrap of a dream. He could barely find Bolivia on a map, but he'd heard that it offered opportunities for industrious Americans. Not long after arriving, he bought a ranch for just over a dollar an acre—and not just any ranch but 37,000 acres of primitive Chiquitano forest and savannah grazing land. With all the U.S. aid coming into the country, the Bolivian government looked favorably upon gringo landholders, and for many decades, Larsen lived a quiet, fulfilling life.

In 2008 that life, as he had come to know it, ended, when he was thrust dramatically onto the political scene by President Evo Morales. Morales had indicated his intention not only to cut ties with the neoliberal economic reforms of the 1990s and the U.S. government but to break up large rural estates. An Aymara "Indian" and Bolivia's first Indigenous president, he made land reform a key principle of his administration.

Larsen insisted that he had acquired his land legally and had treated his workers fairly. The real truth behind Morales's desire to confiscate his land, he believed, was that in 2004 a French energy company had discovered one of the largest natural gas deposits in Bolivia on his ranch. Larsen was convinced the Bolivian government had its sights set on acquiring the gas and that its true intention was to settle Indigenous farmers on the land so that it could negotiate with them. *The New York Times* reported on Larsen's showdown with Morales.

Eventually Larsen lost the battle and was forced to forfeit the bulk of the ranch. Young Duston was allowed to keep his acreage, and his brothers were given the same amount. Ronald Larsen moved to Brazil's Atlantic coast, opened up a bed and breakfast, and penned a book titled *Hijacking My Dream*.

If Duston Larsen still harbored resentment over the government's land seizure, he hid it well. "We're moving on," he said, as the truck climbed the winding dirt lane to the ranch house.

After a late lunch, we took a tour of the land. As we walked toward a rock outcropping where we planned to end the day, Larsen checked his camera traps. I heard him shout "Jackpot!" Sure enough, he'd captured a photo of a pregnant female jaguar, a ranch resident. Everyone was especially excited because females were far more secretive than males and had low detection rates because they avoided trails. It also meant that if all went well, and the cubs were able to escape the high mortality rate of young jaguars, they soon would be romping around the ranch.

After Larsen checked the last of his camera traps, we continued our ascent until we reached a spot that offered a 180-degree panoramic view of the Chiquitano forest and colorful rock ridges that formed the western edge of the Brazilian Shield. It was hard to understand how the jaguar had adapted to this parched and brittle land, dominated by scrub forest with stands of large cacti and palm trees. But according to Rumiz, the forest was brimming with animal life, including six species of wild cat and six species of monkey, marsh deer, white-lipped peccaries, and giant armadillos. Rumiz explained that because of its biodiversity and its connection to the Amazon biome and the Pantanal, the Chiquitano forest of eastern Bolivia was an essential piece of the larger Jaguar Corridor puzzle.

When I noticed a growing hum that sounded like a hive of energetic bees, Larsen told us that we were hearing bats. He called Ezra and me over to examine one of the narrow cracks in the rock wall. I turned on my head lamp and aimed it into the thin fissure. It was packed full of chirping bats that, as the sun faded and fell, were starting to stir. As the light dimmed, the dull din of the waking bats sounded like the approach of a prairie storm. All of a sudden the bats burst from the rock wall. For another ten minutes, we watched and listened and marveled at the dizzying swarm. It was as if they occupied the entire sky.

That night I slept in a hammock on the three-story watchtower behind the house, and at five a.m. I met Larsen at the bottom of the hill, where the ranch hands were just beginning their day's work. Larsen poured me a cup of coffee from a large thermos and then handed my cup to one of the Indigenous ranch hands who, pulling firmly on the teat of a Jafarabadi water buffalo, squirted a long stream of liquid into my cup. When I sipped it, I could taste the milk's richness.

As we watched the ranch hands hobble, milk, and move the enormous buffaloes from the outdoor milking parlor into the corral, Larsen smiled. "It's a great all-round animal. They're very maternal and protective. That's important in jaguar country. When they are grazing, they will always leave one behind to watch over the calves. Then, when it's her turn to eat, another will take her place." Larsen patted the backside of one of the buffaloes. "With the cattle, we have to supplement. The grass here isn't great quality, and it's not enough nutrition for the cattle. But for the buffalo, it is. They're super hardy."

We spent the rest of the day exploring the ranch on foot and by vehicle, and that evening after dinner I briefly toured the ranch's small museum, where Larsen had displayed beautiful pelts of jaguars killed before San Miguelito adopted conservation. Back then Ronald Larsen had a relatively generous three-strikes-and-you're-out policy regarding cattle-killing jaguars. After losing a third cow, he would instruct one of his cowboys to stand watch near the fresh kill and wait for the jaguar to return.

Afterward I sat with Rumiz and Holzmann and discussed Bolivia and its role in a high-profile jaguar trafficking case that reverberated throughout the jaguar conservation world.

Fauna VIVA was a toll-free line, administered by Yandery Kempff, the director of natural resources in Bolivia's Santa Cruz department, to alert authorities about wildlife crimes. On January 16, 2018, the line received an anonymous tip about a sign hanging in front of a restaurant in the city of Santa Cruz: TIGER FANGS ARE PURCHASED. In Bolivia, jaguars are also often referred to as tigers. Kempff's staff secretly visited the restaurant and confirmed the report, setting in motion a potential criminal investigation. A week later Kempff presented the evidence to the police and the environmental prosecutor's office. A Santa Cruz judge then issued a search warrant, which was executed by the Bolivian National Police and the prosecutor assigned to the case. On February 23, 2018, five weeks after the tip, the team took into custody the two owners of the restaurant, Li Ming and his wife Yin Lan. In the process, they confiscated animal parts, skins, jaguar fangs, an unpermitted 22-caliber pistol, and a large sum of money. It was an unprecedented bust for Bolivia. After a preliminary hearing, the judge made the unexpected decision to hold Ming and Lan in the Santa Cruz–Palmasola Rehabilitation Center before their trial. Weeks later the Government of Santa Cruz filed a complaint with the public prosecutor's office for the crimes of destruction and deterioration of state property and national wealth. The law had been on the books for a while but had barely ever been enforced.

Rumiz and Holzmann had been concerned for years about the illegal trade in jaguar parts and Chinese-run smuggling rings. They viewed the Li Ming–Yin Lan arrest as "the tip of the iceberg." Rumiz and Holzmann and other scientists and conservationists in their circle had created a map of Bolivia's protected areas and superimposed upon this map the locations of Chinese companies doing business in Bolivia as part of the Chinese-funded Belt and Road program. They compared this map with government reports and anecdotal information of jaguar deaths and sounded the alarm in Bolivia and across the jaguar range: where Chinese companies were operating, they claimed, there was irrefutable evidence of jaguar body parts being sold for the international wildlife trade. According to data from the national migration office, nearly thirty thousand Chinese citizens entered Bolivia from 2015 to 2017, a period that coincides with China's Belt and Road program, which was especially attractive in Latin America as the Trump administration abdicated the United States' long-standing commitment to the region. China capitalized on the U.S. withdrawal, providing an infusion of money for governments lacking sufficient financing for large development projects.[4]

Rumiz and Holzmann organized a meeting of NGOs and governmental groups to present their stunning evidence. Meanwhile the judge appointed Rumiz and his wife, Kathia Rivero-Guzman, a biologist with the esteemed Noel Kempff Mercado Natural History Museum, and Luis Acosta, associate researcher at the museum, to serve as forensic specialists in the Ming-Lan trial. Rumiz, Rivero-Guzmán, and Acosta spent a month identifying the confiscated animal parts. Their final count revealed that Ming and Lan had in their possession 67 jaguar teeth, which belonged to at least 26 individual cats, 3 jaguar skins, 91 puma teeth from 33 different cats, 8 ocelot teeth from 4 ocelots, a coat made of eight ocelot skins, a peccary tooth, 3 marsh deer skulls, an armadillo claw, 2 rattlesnake tails, and 11 figurines made of ivory.*

By the time the trial began on June 18, 2018, it was a cause célèbre. When the proceedings appeared to stall because of the government's lack of interest in prosecuting a case that had a pronounced anti-Chinese tone and the potential to disrupt Chinese investment in the country, people grew especially vocal. Artists-turned-activists painted murals of jaguars across the city, people protested, and Yandery Kempff joined the demonstrations. Many attended the hearings, showing up with banners. The press too held the government's feet to the fire. The trial was suspended a number of times, but it was forced to resume.

In November 2018, Yin Lan and Li Ming were sentenced to prison, three years for Lan and four for Ming. Lan served her sentence, but Ming, who was freed from preventive detention when the prosecutor failed to appear at his hearing, never served any time.

Nevertheless the seizure and the trial amounted to what was both a national tragedy and a triumph for Bolivia's laws protecting the sanctity of wild animals. The case also served to focus the attention of the Bolivian conservation community on the burgeoning jaguar parts trade.

---

* Rumiz, Rivero-Guzmán, and Acosta wrote guides to wild cat parts to help officials in Bolivia, Brazil, Peru, Colombia, and Suriname know what to look for.

CHAPTER 16

Land of Thirst

AFTER MY CONVERSATION WITH DAMIÁN RUMIZ AND ANAI Holzmann, jaguar poaching, and its growing prevalence, was on my mind. The Lan-Ming case had been a success, but jaguars across Latin America, especially in Bolivia, were still being killed at alarming rates.

The filmmaker Elizabeth Unger knew this as well as anyone. In her riveting and award-winning documentary *Tigre Gente*, she followed Marcos Uzquiano, the park director for Madidi National Park, who had dedicated his life to protecting jaguars, as he attempted to penetrate a network of hunters and traffickers. While Unger accompanied Uzquiano through the thick forests of Madidi, her film tracked the Hong Kong journalist Laurel Chor. Chor was investigating the selling of jaguar teeth in China and Myanmar and trying to establish a link between the poaching of big cats in Latin America and Chinese demand for jaguar parts.

"Nobody else in the conservation world was really talking about it," Unger said. "I heard what was happening in Africa, but I had never heard of jaguar parts being trafficked for the Chinese black market.... We followed [Uzquiano] for three years. I wanted to show what it was like for the rangers."

Despite the best efforts of supremely dedicated people like

Uzquiano, Bolivia's jaguars, even in sanctuaries like Madidi, were in peril. Kaa-Iya del Gran Chaco National Park may have been one of the last places in the country where jaguars were relatively safe.

At 8.5 million acres, almost twice as big as Madidi National Park and four times the size of Yellowstone National Park, Kaa-Iya is the largest protected area in Bolivia. Straddling the border between Bolivia and Paraguay, it is part of the most expansive dry tropical woodland forest—called the Gran Chaco—in the world. The entire Gran Chaco, which once spread over a huge area, from Argentina to Paraguay to Bolivia, now ranks as the number-one deforestation hotspot on the planet. It is more imperiled even than the Amazon. Between 1987 and 2012, the country of Paraguay alone cleared nearly 11 million acres of Gran Chaco forest to make room for cattle ranches and soybean fields as big as small cities and for the production of charcoal.

Kaa-Iya is deceptively significant. Although the Gran Chaco forest, in general, looks almost incapable of nurturing life, much less charismatic megafauna species, it is prime jaguar habitat. The word *chaco*, in fact, stems from the Quechua word *chaku*, meaning "hunting land," a testament to its diversity of wildlife. Though Kaa-Iya gets only sixteen inches of rain per year, it has exceptionally high levels of mammalian species diversity. It is a place, too, that held a special allure for Alan Rabinowitz, in part because of its extraordinary and unexpected biodiversity and because Rabinowitz was attracted to wilderness on a grand scale.

Despite its many attractions—tapirs, the elusive maned wolf, the largest canid in South America, pumas, ocelots, and jaguars—Kaa-Iya sees few visitors. The park is only 130 miles from Santa Cruz, and while tourists do visit the small town of San José de Chiquitos, a former Jesuit mission and now a World Heritage site, few venture into the country's barren and obscure south. Those who come to Bolivia usually visit only a handful of places, including the more famous Madidi National Park, far to the north. Madidi is classic jungle terrain, and it's mountainous, whereas Kaa-Iya, despite the melody of its name, is flat, monotonous, almost haunting in its austerity, and chock full of thorns and stunted trees. Its weather is dominated by extremes. After the austral winter, it smolders with heat. Temperatures frequently rise above 100 degrees Fahrenheit. But in winter the temperatures can hover near freezing. Water is scarce.

To the Indigenous Guaraní, Kaa-Iya has always been a special and sacred place. In their language Kaa-Iya means "spirit guardians of the forest." Isoseño-Guaraní people live on the park's western edge along the Parapetí River, while Mennonites farm land to the northwest. In the south, a nomadic, hunter-gatherer people called the Ayoreo consider themselves the "true people" and reject contact with the outside world. They occasionally leave behind beguiling evidence of their presence but are unwelcoming, remote, and wild.

In 1932 a war broke out between Bolivia and Paraguay over ownership of the northern part of the Gran Chaco. Some referred to the war poetically as La Guerra de la Sed (The War of Thirst) because it was fought in one of the most arid regions in all of South America. Indigenous Isoseño-Guaraní were sandwiched between the two opposing armies. When the war ended in a pyrrhic victory for Paraguay, the Isoseño-Guaraní people initiated a campaign to gain title to their lands, which culminated in 1995 with the creation of the Kaa-Iya National Park, a milestone for Bolivia's Indigenous peoples. Extending over 8.4 million acres, it was one of the first protected areas in South America to be established through the initiative of an Indigenous group.*

---

* In 1991 Capitanía del Alto y Bajo Izozog (CABI), a grassroots Indigenous organization, enlisted the support of the WCS in its efforts to legitimize its territorial land claims. The collaboration was based on the twin principles of sustainable use of renewable resources and conservation of the region's rich biodiversity. The two groups began exploring the possibility of creating a national protected area in the Gran Chaco. In 1993 that collaboration led to the creation of the Kaa-Iya del Gran Chaco National Park. The Bolivian government approved the proposal in 1995, then awarded CABI responsibility for co-managing the park. Working with the WCS again, CABI implemented an environmental education curriculum that taught children about Isoseño-Guaraní history, traditions, and culture.

For the WCS, it represented an opportunity for the NGO to play a pivotal role in Bolivia's environmental future. That role would be challenged when the gas line was being promoted, but CABI and the WCS were invited to participate in the process and comment on the environmental impact statement as well as on the route of the pipeline and its design, with an eye toward reducing the short-term and long-term impacts of the pipeline and the inevitable changes to the regional economy that hydrocarbon development would bring.

However, two years after the park came into being, the newly acquired rights of the Isoseño-Guaraní were challenged, as construction crews descended on the park to begin building a corridor to accommodate the $2.2 billion Bolivia-Brazil gas pipeline, the largest energy sector project ever undertaken in South America, which would connect the gas fields of southern Bolivia with the Brazilian cities of São Paulo and Porto Alegre. Many Isoseño-Guaraní feared the worst. But Petrobras, Brazil's state-run oil company, showed that it was willing to accommodate their concerns. To this day, conservationists consider the gas line a model infrastructure development project and praise it not only for its high environmental standards but also for the way it handled and honored Indigenous land claims.

TO REACH KAA-IYA, Ezra, Saul, our irrepressibly energetic biologist-guide, and I drove south of San José on a two-lane highway, then turned west onto a long dirt road that transected the park's buffer zone. Not long after we'd crossed into the buffer zone, we saw a man walking with a shotgun slung casually over his shoulder. As we passed, he smiled and waved, unconcerned that hunting in the buffer zone, as the occasional NO CAZAR (no hunting) signs announced, was explicitly forbidden.

We traveled slowly, watching for animals along the way, and used the time to get to know Saul, who as a young Bolivian man had received a scholarship to study at Cambridge University, where he'd developed a love for British eccentricity, fish and chips, porridge, Guinness, and haggis. He peppered his speech with distinctly British words and phrases, like *fantastic* and *bloody hell*, even nailing the inflection. After England, he went to Montana State University for his master's in wildlife management and while there spent six months in Yellowstone studying wolves. He enjoyed the United States, especially his work in Yellowstone, but he reserved a special affection for England and most things English.

As we drove through the buffer zone, the air hung heavy with smoke. In 2019 fires in Bolivia scorched 6 million acres. In 2020

Bolivians hoped for a reprieve, but what they got were skies choked with smoke, ash storms, and furnace-like fires that torched another 4 million acres. In 2021 an additional 2.5 million acres burned.

As we drove farther west, the skies began to clear ever so gradually. The sun, tinted almost crimson by the smoke, was beginning to set when we arrived at the park gate, where the rangers informed us we were the park's only visitors. Ezra and I joked that we'd never had a national park all to ourselves. Nick McPhee, always the rebel, would say later that despite what we'd heard about the wonders of the gas line, the park's directors, political appointees, discouraged tourism and had little understanding of what Kaa-Iya represents or what it could be. For the most part, he said, they were there to appease the gas company.

We quickly unloaded our gear into the station's very basic guest quarters. We soon learned we would be sharing it with two large tarantulas, who spent their time hiding in the corners and scurrying across the floor. Then we returned to the truck for a night of wildlife viewing. Saul told us that once the sun set, we should look for the jaguar's mesmerizing blue-green eyes along the side of the road. He imitated the sound of a roaring jaguar, a gruff croaking and coughing that he'd heard one night and would never forget. Because they possess a modified hyoid bone in their throat, jaguars can roar. In fact, they are the only cats in the Western Hemisphere that can. They can roar, grunt, woof, growl, and moan and so communicate with other cats over large distances. Some biologists say that a jaguar's call is its acoustic fingerprint, as singular as its rosettes.

Our plan for the evening was to drive until midnight and then wake two and a half hours later in order to be back on the road between three and six a.m., when the camera traps that WCS biologists were maintaining indicated that jaguars were most often on the move.

We headed off on a long stretch of road, lit by emerging stars, that felt like it hadn't been graded in ages. Dodging ruts and potholes, Saul said the road followed the route of the gas line. Depending on the weather, he said we might make an attempt the following day to cover the arduous sixty-two miles, ending near the Isoso Camp, in the park's northwestern corner.

Saul told us, too, about a multiyear study of the park's jaguars by the WCS that concluded that Kaa-Iya's big cats had adapted to the gas line—a thirty-two-inch diameter pipe that ran for eighty-one miles across Kaa-Iya's landscape and required fifty feet of clearance on either side—and the pumping stations and the occasional noise of trucks and graders that came with running and maintaining dirt access roads, compression stations, and generators.

The heroine of that WCS study was a remarkable and self-possessed female jaguar that biologists had named, appropriately, Kaaiyana. She had clearly charmed both the WCS biologists and the park's guards, who saw her regularly over a six-year period and occasionally observed her with a mate and, later, with cubs. Kaaiyana seemed not to be disturbed by the gas pipeline or the concomitant activity and was unusually calm and unaffected by the presence of people. She would lie in the middle of the road, just feet from the vehicles of the biologists or rangers, with her cubs beside her, and sometimes she would even nurse them unwarily. Kaaiyana exhibited what biologists call "range fidelity." The park's male jaguars roamed, especially in the midst of the dry season, when jaguars in general move over greater distances in search of water, but Kaaiyana was dedicated to her particular part of the park.[1]

That first night we saw a brocket deer, a tapir, and what looked from a distance to be a big cat crossing the road. When it disappeared into a thicket, we parked the truck, turned on our flashlights, and set out on foot in hopes of finding tracks. Ezra discovered them, and Saul confirmed that it was a puma and not a jaguar. Shining his light on the imprint, he said that a jaguar's tracks were larger and wider, with toes spread farther apart than a puma's. The puma's, by contrast, were more streamlined.

By the time we returned to the ranger station, it was almost midnight. As tired as I was, I stood outside the station building and marveled at the beaming stars, which seemed to have overtaken the entire sky. When I went in, the air was still heavy and hot, and I slept little. When I finally did doze off, I woke with a start, convinced that a tarantula had just scuttled across my bare chest. By the time Saul knocked on the door at two-thirty to wake us, I was already in the kitchen, having grown accustomed to the near-sleepless schedule that these trips entailed.

When we stepped outside, a light rain greeted us. Saul decided it would be wise to check some of the camera traps alongside the road, and watch the weather, before heading for the Isoso Camp. After spending many weeks in the field, in a variety of different countries, I'd grown fond of the ritual of checking camera traps and the unexpected surprises they sometimes offered.

The first camera revealed two tapirs, a mother and a baby; the second, to Saul's delight, a jaguar cub; and the third, a large male puma and a female. At first light, not long after checking the final camera, Ezra spotted a solitary sandpiper on the side of the road. By his reaction, I would have thought he'd seen a jaguar. But for Ezra, a lifelong birder, seeing a sandpiper in October in Bolivia was the equivalent of spotting a wild cat. He explained that the birds were world-class travelers and that this one had likely just arrived from Alaska or Canada via the Mississippi Flyway.

Although we had set our sights on the Isoso Camp, by dawn we realized that Mother Nature wasn't cooperating. A warm, steady rain fell, and the road became as slick as grease and filled with water, making a mockery of our plans.

By daylight, our four-wheel-drive truck was spinning and weaving, and Saul was barely keeping it on the road. At one point, he lost complete control, and after doing a full 360, the truck ended up on the side of the road mired in mud. Ezra and I got out and pushed mightily until the truck began moving again. We ran along, slipping and sliding, then jumped in like bandits running from the scene of a crime. I'd just situated myself in the front seat when the truck began to spin again, ending up in another shallow ditch. Ezra and I waded back out into the morass. But this time, when Saul stepped carefully on the accelerator, and we pushed with all our weight, the wheels simply spun in place. The truck wouldn't budge.

Forced to admit that we were stuck, we realized we were thirty miles from the ranger station with no way of alerting anyone that we might need help. The spinning tires had created deep, waterlogged ruts. The rain was falling harder than ever, and we decided to try to wait it out.

What the rain gave us was time to talk. Saul spoke optimistically about jaguar populations in the Gran Chaco region of Bolivia and Paraguay. He explained how just two years after Bolivia created Kaa-Iya, it established the Otuquis National Park and Integrated Management Natural Area, east of Kaa-Iya. It too was large—almost 2.5 million acres. Another addition to the Gran Chaco ecosystem, the Ñembi Guasu Area of Conservation and Ecological Importance, occupied the territory separating Kaa-iya and Otuquis. Formed in 2019, it protected nearly 3 million acres.

In the Guaraní language, Ñembi Guasu means "the great hideout" or "the great refuge," which is apt considering it would help to defend a vast connected ecosystem. Together Kaa-Iya, Ñembi Guasu, and Otuquis form an enormous safety zone, where large mammals can move freely and flourish. When Alan Rabinowitz conceived of the idea of Jaguar Conservation Units, functioning as repositories of biodiversity, with large numbers of jaguars, some of which would branch out into new territories to invigorate the gene pool, this is exactly what he was thinking of.

It is one of the few places in Bolivia where ambitious long-term plans can be made for growing the jaguar population. At almost 14 million acres, it forms a continuous conservation area, a vital corridor extending all the way to the Brazilian Pantanal. Jaguar biologists conjecture that the wilderness area, the Earth's largest protected dry forest, may contain more than one thousand jaguars.[2]

As heartened as Saul was by this achievement, he worried about Bolivia's political turmoil as well as the deleterious effects of the pandemic. He confessed that Bolivia, in many ways, was a country at odds with itself, struggling to honor Indigenous land claims and to protect the jaguar and its habitat, while rushing headlong into the modern agricultural economy. The pandemic had exacerbated the divide. Ecotourism operations, like San Miguelito, that offered environmentally friendly alternatives to logging, poaching, and large-scale agriculture, were faltering. Even Saul, a trained biologist who had tried to carve out a living as a guide in Bolivia's fragile and fledgling tourist industry, was struggling. But it was Bolivia's poor who were really suffering. People who had lost their jobs due to the pandemic and the accompanying economic downturn had resorted to supplying the burgeoning wildlife trade with jaguar parts. "Money has power," Saul said, as if to explain why the laws on the books designed to protect the jaguar could not stand up to the grim realities of poverty.

A recent study by a group of conservationists from the WCS, Cornell University, Universidad del Pacífico in Lima, Peru, and Zamorano University in Honduras underscored Saul's fears. The first comprehensive review of national laws across the jaguar range, it revealed the strengths and shortcomings of wildlife laws and the difficulties of enforcing legal penalties when they are broken. In Mexico, for instance, the illegal hunting of jaguars officially brought a fine of up to $500,000 or three years in prison, but the penalty, emblematic of what was happening up and down the corridor, had never been enforced, not even once.

As our conversation trailed off, the rain fell prodigiously, and by early afternoon any hope we had of reaching the Isoso Camp had clearly been extinguished. Now we were concerned only with making it back to the ranger station. During a brief break in the deluge, we left the truck to reconnoiter, entertaining the notion of trying to walk back. But in the slick mud and clay, we could barely stay on our feet, and in some places along the road, the water had gathered into waist-deep pools that proved almost impassable. In fact, water had penetrated everything, and dry ground had all but disappeared. That's when we realized that hiking back to the ranger station was in all likelihood the most difficult and dangerous of our options. Thirty miles, under the best conditions, was a very long hike. And what kinds of venomous spiders and snakes—Kaa-Iya is a hotspot for neotropical rattlesnakes, coral snakes, and pit vipers—were slithering through the murky pools? The thought made us shiver with fear.

We spent the next few hours dozing and talking, and by late afternoon, the rain suddenly stopped. We emerged, aware that we would need to rescue ourselves. So we gathered rocks and armfuls of branches, filled the deep ruts as best we could, and chocked the tires.

Pushing with all our might, Ezra and I rocked the truck back and forth for several minutes as Saul pressed on the accelerator. Just as we were about to abandon our efforts, the truck jumped out of the ditch with the tires spitting mud. Before we knew it, Saul was weaving down the road. By the time Ezra jumped in, I was face down in a wallow of mud. I worried they would just keep going. But Saul slowed, while I spat the mud from my mouth and ran to catch up.

Wet and wheezing, I jumped onto the running board and opened the door. We had succeeded in getting the truck back on the road, just in time. From my seat, I could see the day's last light lingering along the horizon.

As Saul turned on the headlights, he told us to wind up our windows quickly. We couldn't understand the urgency until we saw them: clouds of frenzied moths, termites, and insects stirred up by the incessant rains swarmed around our headlights.

"Bloody fantastic!" Saul shrieked.

Then something hit the windshield like a wet sponge.

"What is it?" I asked, as I heard the sound again and again.

Saul didn't even bother answering. He just smiled. When I shined my headlamp, I saw what was colliding with our windshield. Frogs! They were leaping from the trees onto the truck in a kind of ecstatic celebration. Their slimy mucus was splattered across the glass. But they weren't dead. They attached themselves to the windshield, hitched a ride, then jumped off.

Trapped in a kind of phantasm, we kept driving in the pitch-black darkness, as the crazed frogs kept jumping, and bugs filled the sky. The rain had transformed the park's forest into a wetland, and the road resembled a swamp fit for a U.S. Army amphibious vehicle. That Saul was able to keep the truck moving in the right direction was a miracle. Seven hours later we reached the ranger station.

We slept little that night and were out early, well before sunrise, because we knew animals would be hunting and foraging after the rain. By eight a.m., after four hours of searching, Ezra resigned himself to the probability that, once again, we would not see a jaguar. His regret reminded me of a passage in Peter Matthiessen's *The Snow Leopard*, in which Matthiessen joins George Schaller tracking the secretive and solitary cat in the Himalayas; they never find one. "I am disappointed," Matthiessen confesses to Schaller, then adds, "And also I'm not disappointed." Schaller then muses, "We've seen so much. Maybe it's better if there are some things we don't see."[3]

In Kaa-Iya alone, Ezra and I had seen a shadowy panther walk across the road and two tapirs. We'd experienced a breathtaking outburst of life, the sky literally raining frogs. And on that last morning, we saw two tayras and what might have been a jaguarundi. The tayras moved back and forth across the road, pouncing ravenously on frogs and moths that had been attracted to the temporary puddles.

Ezra had interrupted his work with his beloved orcas in the Salish Sea to come to Bolivia. Although we never got a glimpse of a big cat, he had added 167 new bird species to his life list, which he had been compiling since he was seven years old. He too was awestruck by Kaa-Iya and by the possibility that somewhere in those stunted scrub forests, the shining green eyes of the jaguar had been watching us.

CHAPTER 17

Field of Dreams

AT THE END OF 2001, RABINOWITZ WAS ANGUISHING OVER the implications of his disease. On the one hand, he wanted to spend his remaining years with Salisa and his son Alexander—fatherhood, he wrote, had been a "blessing"—but on the other, he dreamed of returning to the field to do the kind of work he'd always loved.[1] Could he do so without adversely affecting his health? He got his answer when, thanks to Jane Alexander, he met with the chief of hematologic oncology at the Memorial Sloan-Kettering Cancer Center. Dr. Stephen Nimer assured him that he was in the earliest stages of the disease and would be around for a while, assuming he took care of himself. Nimer told him to "be careful." If he contracted an illness in the field, it could "kick [his] immune system into high gear and speed up the disease."[2]

Rabinowitz spent the next few months fighting off depression. But then in May 2002 his second child, Alana Jane Rabinowitz, was born—the name Alana for him, and Jane for her godmother Jane Alexander, Alan's trusted friend, confidante, and travel partner for almost two decades. Curiously, Alana's birth seemed to right Rabinowitz's faltering ship.

Half a year later, reinvigorated, he was back in Myanmar as global politics infringed on his conservation plans for the country. The United States had suddenly become emphatic about isolating Myanmar, pressing for sanctions and relegating it to international pariah status. Fearing some kind of governmental reprimand, Rabinowitz hurriedly returned to the United States and decided to attempt something drastic from the security of his home: in the interest of the animals, he would blatantly ignore politics and write General Khin Nyunt, head of military intelligence, one of the most internationally reviled men in Myanmar's regime, and appeal to him on behalf of the Hukawng Valley.

One year later came the stunning news. General Nyunt had authorized the head of the forest department to set aside 8,500 square miles as the Hukawng Valley Wildlife Sanctuary. The reserve was, in Rabinowitz's words, "one of the most complex protected areas ever designed," a tapestry of tiger habitat and small communities that would serve as a new model for the kind of visionary conservation projects Rabinowitz believed large predators required.[3]

In 2006 Rabinowitz left Myanmar for good and returned triumphantly to New York, having accomplished what even he thought was the impossible. He had done everything in his power for the tiger, yet he was painfully aware that that conservation groups were "losing the battle for the species."[4] He would later write, gloomily, in *An Indomitable Beast*, that the tiger, despite being "one of the most iconic and recognizable animal species on the planet" and despite all the efforts on its behalf, was sliding swiftly "towards extinction."[5]

If after his cancer diagnosis he was uncertain about his ability to continue his work on behalf of big cats, he no longer doubted his mission. His goal now was to do for the jaguar what he had not been able to achieve for the tiger. The difference was that the condition of the tiger, which occupied only 7 percent of its historic range, was "akin to a body on life support in the intensive care unit of a hospital."[6] The jaguar, however, still stood a fighting chance, and although he would "never abandon the fight for the tiger," the jaguar was "still the animal closest to [his] heart."[7]

CHAPTER 18

The Jaguar Corridor

BACK IN NEW YORK, RABINOWITZ PUT TOGETHER HIS core jaguar team, a portion of which he had begun assembling not long after the pivotal meeting at Cocoyoc. Its first task was to identify where jaguars were and where they weren't by gathering data from 110 jaguar experts across Central and South America. Eric Sanderson, a landscape ecologist and a recent WCS hire, was in charge of assembling and organizing the data, which he superimposed over a map of the known jaguar range. For the young Sanderson, not long out of grad school, it was an intimidating assignment.

"Here I was working on a project of enormous magnitude with these scientists who were the best and brightest in their field," Sanderson says. "Alan didn't fit the mold, or at least my image of what a wildlife biologist was a supposed to look like. He was a strong, forceful man, and already a legend in his field. But he was respectful of my opinions."

Sanderson rose to the occasion. As lead author, he had penned an influential paper in *Conservation Biology* that urged conservationists hoping to save "broadly distributed species . . . to plan explicitly for their survival across their entire geographic range and through political boundaries."[1]

To assist Sanderson, Rabinowitz brought abroad another landscape ecologist, Kathy Zeller, a talented GIS (geographic information systems) expert, who had done her master's on large landscape conservation networks. Upon graduating, she wrote to the WCS. Not long afterward she found herself sitting in Rabinowitz's office at the WCS's New York headquarters at the Bronx Zoo. Although she knew of his reputation, she was still surprised by how his aura filled the room. After explaining his idea for the Jaguar Corridor, he asked her point-blank if she had the skills to accomplish the kind of mapping he needed. Zeller answered bravely that she thought she could do it.

Zeller updated the mapping effort and added the potential corridors to Sanderson's model. She mapped pathways linking jaguar populations, taking into account proximity to water, distance from roads, towns and cities, elevation, and the presence of vegetation. The map was not a theoretical model; rather, it reflected a pathway, or a series of stepping-stones, on a country-by-country basis that the jaguar was likely using.*

What really got the fledgling corridor off the ground and onto the conservation map was the support of Costa Rica's minister of environment and energy, Carlos Manuel Rodríguez-Echandi, a friend and admirer of Rabinowitz and an ardent conservationist and jaguar lover.

In 1998, not long after he was appointed Costa Rica's minister of the environment, he read *Jaguar* and was impressed that a biologist from the United States would risk his life to study Central America's premier terrestrial predator. Rodríguez promised himself that one day he would travel to New York to meet him. In 2003 he made good on that promise, and the two established an immediate rapport.

In March 2006, Rodríguez invited his friend to present his Jaguar Corridor Initiative to a gathering of all the ministers of the environment from Central America, Mexico, and the Dominican Republic. Rabinowitz knew he would need to win over their hearts and minds,

---

* In 2006 Zeller condensed all the surveys, graphs, and numbers into a central report for WCS that would guide jaguar recovery efforts for well over a decade. Zeller, *Jaguars in New Millennium*, 1–82.

channeling the kind of persuasive energy he had summoned more than two decades before, when he spoke to the prime minister of Belize and his cabinet about the necessity of preserving Cockscomb. It was a daunting setting, a collection of influential ministers whose imprimatur he needed for the corridor to succeed.

Rodríguez introduced Rabinowitz as a "beacon of light."[2] In his speech, Rabinowitz returned the compliment, telling the assembled ministers that in Rodríguez, he saw a conservationist capable of uniting Central America around a common environmental goal with the jaguar as its signature species. He described the growing threats to jaguar populations. Then he divulged the genetic discovery—that there was only "a single taxon"—and his proposition for maintaining those genetic links throughout the jaguar range. Kathy Zeller provided him with the maps he needed to illustrate his concept of the corridor, enabling the assembled ministers to visualize the pathways and their effect on jaguar dispersal.[3]

After the speech, Rodríguez enthusiastically committed Costa Rica to backing the Jaguar Corridor. Thirty minutes later the delegation voted unanimously to support the corridor. Though heartened by their enthusiasm, Rabinowitz knew that the hard work of physically verifying the corridors would require an enormous, range-wide effort. This would ultimately take five arduous years to accomplish.

Roberto Salom-Pérez was in charge of the all-important ground-truthing effort. Salom-Pérez took the Jaguar Corridor Initiative on the road, beginning in Costa Rica, Nicaragua, and Honduras, then moving on to the rest of Central America. Along the way, he used his considerable skills to recruit and train other native scientists. Working to make the corridor a reality, they adjusted the computer model to fit field assessments gained from camera trap surveys and thorough site searches. They also conducted interviews with rural communities, ranchers, and in some cases even poachers.*

---

* Salom-Pérez would always be proud of Costa Rica's role as an early corridor model. The first approved corridor was the Talamanca-Central Volcanic Region, which after ground-truthing efforts, was moved west of where the initial map indicated.

When Salom-Pérez and his crew of researchers finished, Zeller went back to work, ranking the corridors on a 0–10 basis, with 0 representing a landscape where jaguar movement was unimpeded by natural and man-made obstacles, and 10 indicating one that had been seriously degraded. What the team discovered, to its great surprise, was that despite the substantial loss of habitat, burgeoning roads, cities, and expanding agricultural fields, the ever-adaptable jaguar was using 80 percent of its historic range.

Next, Rabinowitz and his team returned to the concept of Jaguar Conservation Units. Using data submitted by jaguar experts, they mapped and ranked the JCUs, which represented three-quarters of a million square miles and a surprising 16 percent of the entire jaguar range. Type 1 JCUs they defined as thriving units with at least fifty breeding jaguars and an abundance of prey, while Type 2 units had ample prey and favorable but imperiled habitat and fewer than fifty jaguars. Next, they mapped the corridors connecting ninety JCUs, arriving at a sum total of 182 corridors, comprising one million square miles, spread across eighteen countries and two continents. Of the 182 possible corridors, Zeller discovered that only 46 percent were protected and almost a quarter of them were less than six miles wide and fragile enough that they could easily be lost to development. Zeller warned that losing any one of them could sever the trans-American jaguar population.

When Rabinowitz described the corridors and what they meant to the jaguar's survival, the analogy he used was the Underground Railroad, the network of secret routes and safe houses established in the United States that helped an estimated 100,000 enslaved people escape to free states in the North and to Canada. The essence of his argument was that migration and safe passageways were the keys to genetic freedom. Essential also to this analogy was the groundbreaking work of E. O. Wilson and R. H. MacArthur. The two had jointly authored *The Theory of Island Biogeography*, the first book to advance the idea that islands lose biological diversity at especially high rates. Fragmentation of habitat, they wrote, was biodiversity's worst enemy because it brought about genetic isolation. The basic message was that islands were where species go to die.

Rabinowitz was also influenced by his friends Howard Quigley and Peter Crawshaw and their 1992 paper on how to save the jaguar in the Brazilian Pantanal. They wrote that corridors were "preferable to isolated refuges" regardless of their size. The appearance of one new jaguar, introducing fresh DNA to a group of fifty jaguars, would fortify a population, providing it with a better chance of survival than an isolated island with one hundred jaguars.[4]

Rabinowitz came to embrace the groundbreaking idea that turned on its head the prevailing conservation paradigm, envisioning the jaguar realm as a mammal's circulatory system. The JCUs, which functioned as the core areas of jaguar production, were its heart, while the corridors were its veins and arteries. A functioning system would nurture the species while at the same time allow individual cats to spread their genetics across the corridor.

At some point it dawned on Rabinowitz that he was attempting to implement perhaps the largest and boldest conservation model in history. Never one to shrink from a challenge, he did the only thing he knew how to do. Despite his cancer diagnosis, he committed himself heart and soul to protecting the jaguar's future. That meant using his considerable power of personality to convince governments to formally pledge their cooperation in developing "national jaguar action plans," which committed them to strict "guidelines for protecting jaguars, monitoring the jaguar corridor, training personnel, and addressing jaguar conflict issues."[5] It also required that governments assimilate and use corridor data to inform their future zoning plans and to consent to performing environmental impact statements on all future development projects that might impact the corridor. Those commitments took the form of Memoranda of Understanding (MOUs).

Acquiring those signatures required a kind of high-stakes diplomacy, and Rabinowitz was now a seasoned pro. In February 2010, Colombia's vice president Francisco Santos, the minister of the environment, and the director of the national parks invited Rabinowitz to meet them in Bogotá. Remarkably, Colombia became the first country to sign an MOU, followed by Costa Rica. For Rabinowitz, it was an extraordinary achievement especially because he and Kathy Zeller considered Colombia's Darién Gap to be one of the most vulnerable corridors of the entire initiative.

But coaxing governments to embrace an unprecedented land-use model that stressed the importance of conservation amounted to winning only half the battle. Rabinowitz needed to form similar partnerships with "local officials, landowners, protected area managers, local businesses and developers, farmers and ranchers, teachers, and indigenous groups" for whom economic opportunities were often scarce.[6] They would be the ones whose lives were most impacted. These were the same people to whom the short-term rewards offered by "open-pit mines, intensive forestry projects, large mono-culture plantings, or hydroelectric dams" were often directed. He needed to create alternatives that fostered a sense of "ownership and pride in their natural resources," and when that failed, he had to be willing to work with companies to develop effective management strategies that would reduce impacts on wildlife.

Like Schaller, he realized that "principles must always be juggled with practicalities."[7] Schaller wrote, "A field biologist must work with local people . . . to find innovative solutions. The establishment of a nature reserve often creates hostility, for, deprived of their land, local people are no longer able to collect fuel, hunt, and graze livestock. The traditional lives of indigenous people must be considered. . . . One cannot discuss the philosophy of conservation with a man cutting the last tree for fuel." For the Jaguar Corridor to succeed, Rabinowitz would need to solicit the participation of vastly different peoples, representing dramatically different cultures, spread across an awesome five thousand miles of terrain.

Rabinowitz's team felt a profound sense of urgency. Landscape fragmentation was occurring at a breakneck pace, and key corridors, which were often multiuse areas with minimal protection, were disappearing before conservation efforts could save them. What would ultimately preserve the integrity of the entire corridor rested on the team's ability to secure these lands by tapping into what Rabinowitz had identified as the Jaguar Cultural Corridor, where the mystique of the big cat still existed and where the jaguar was still "part of the life of the people among whom it lived."[8]

For a long time, Rabinowitz had viewed the people who lived among the jaguars mostly as "antagonists to the animal's survival." They developed the jaguars' land, hunted their prey, shot them for trophies, and killed them out of fear. Their historical and cultural connection was "part of a dying past that no longer had a place in the modern world." But the more he learned about the people of Central and South America, the more he came to understand that he had been wrong.[9]

CHAPTER 19

The Billionaire Who Loved Big Cats

IN PURSUIT OF HIS VISION OF THE JAGUAR CORRIDOR, RABinowitz eventually partnered with Tom Kaplan. Kaplan was a polymath, a scholar with a love of history, art, and the natural world, and a billionaire unapologetic about the fortune he had made in precious metals. But as passionate as Kaplan was about silver and platinum, what he really wanted to do was to protect wild cats, especially the six big cat species—lions, tigers, jaguars, leopards, cheetahs, and pumas.

As a child growing up in New York City and then Pompano Beach and Fort Lauderdale, Florida, Kaplan spent considerable time in the outdoors. He was also a voracious reader, and an article about tigers, written by George Schaller, inspired his interest in felid zoology.

At age eleven, he accompanied his mother on a trip to the Tres Fronteras region of the Amazon to search for jaguars, all the while dreaming of becoming a wildlife biologist. Instead, he would go on to earn a Ph.D. in history from Oxford University, where he had also acquired his bachelor's and master's, titling his dissertation "In the Front Line of the Cold War: Britain, Malaya and South-East Asian

Security 1948–1955." He might have gone on to inhabit the lecture hall as a professor, but at that pivotal moment, he chose to pursue a career as an entrepreneur. He began working for the Israeli investor and hedge funder Avi Tiomkin as a junior partner, focusing on currency markets. But what really fascinated him was silver, which was trading at around three dollars and fifty cents an ounce. Kaplan believed it was dramatically undervalued. He looked for an institutional-quality silver company to recommend to Tiomkin but couldn't find one and contemplated starting his own company. When he told his wife, Daphne Recanati, the daughter of prominent Israeli banker Leon Recanati, what he was thinking, she wholeheartedly supported him.* In 1993, with seed money from George and Paul Soros (his mentor) and Jack Nash and the entirety of his own savings, he took a gamble and started Apex Silver. And then he watched the price of silver rise. He ended up making 220 times his investment and transitioned to platinum in Zimbabwe and South Africa, where he made over 100 times his investment. Following that, he put his money into hydrocarbons in Texas, where again he made a killing.

Thanks to his keen instincts, intellect, chutzpah, and what he called "La Fortuna," he made billions. But gradually his interest in wildlife conservation took center stage. He'd always admired visionaries such as Ted Turner, and Doug and Kris Tompkins, who managed to turn their environmental consciousness and nearly limitless capital into epic conservation initiatives.

AFTER STARTING PANTHERA IN 2006, Kaplan's initial goal was to work in concert with the WCS, where Rabinowitz was the executive director of science and exploration and the head of the big cat program, and other organizations as "partners in wild cat conservation." He had thought the world of conservation was based on a "Cumbaya spirit of good will," but when he realized how naïve he had been, he abandoned the notion of comity and cooperation and changed Panthera's motto

---

* Kaplan says that his wife, Daphne, called it "sexy."

to "Leaders in Wild Cat Conservation." The "institutional divorce," as one WCS employee called it, created hard feelings.

Then Kaplan began courting Rabinowitz, who was intrigued but skeptical. It seemed imprudent for a scientist in his fifties, dealing with chronic lymphocytic leukemia, to leave a respected NGO like the WCS for a conservation start-up. The disease could rear its scary head at any moment. Moreover, he had a wife and two children. But Rabinowitz was not a timid man, and he was increasingly persuaded by Kaplan's pitch: Panthera was going to be a lean, mean organization, but one with the money to make a difference. Rabinowitz was aware of the stark statistic: Only 2 percent of global philanthropy went to the environment, and of that, species conservation got only a fraction, which meant that wildlife conservation groups were perpetually hamstrung by a lack of cash. Kaplan convinced him that Panthera would be different. It would have a mission and the money to fund it.

As painful as it was for the WCS to lose someone of Rabinowitz's stature, not everyone was unhappy to see him leave. He was brilliant and utterly committed to protecting big cats, but he could also be hardheaded and narcissistic. Some thought his fascination with the corridor was misguided. They worried about the vitality of the Jaguar Conservation Units, about pinch points developing in what should have been the species's strongholds. Without stable JCUs, there would be nothing to connect to. Many were also puzzled by the unusual relationship—the world's foremost big cat biologist teaming up with a billionaire many called "Gold Bug" who had made his fortune by tearing open the earth.[*]

For Kaplan, bringing Rabinowitz aboard to head up his fledgling organization was like striking, simultaneously, silver and gold. Rabinowitz was the "greatest brand name in wildcat conservation." Kaplan called him "the charismatic megafauna of his generation." As for the Jaguar Corridor, Kaplan, who had always been "attracted to the magnificent," was dazzled by its scope. He had long felt a sense of urgency about extinction, not a complete blinking out, like the pas-

---

[*] Kaplan's Electrum Group owned (and still owns) a diversified international portfolio of gold and other precious and base metals like copper, nickel, and zinc.

senger pigeon, but a fragmentation of habitat so severe that genetic transference was no longer possible. The concept of the corridor meant that Panthera could embrace a compelling model of conservation that allowed it to avoid what he called "the trench warfare" waged by so many other conservation groups. Best of all, it was a model based on an "already existing corridor" created by the "tenacity and resiliency" of the far-ranging jaguar.[1]

If at any point Rabinowitz had been ambivalent about leaving the WCS, his first trip to the Brazilian Pantanal with Kaplan provided all the convincing he needed. After Ricardo Boulhosa, a WCS field biologist working in the Pantanal, convinced Kaplan that the region offered the best chance of seeing a jaguar in the wild, Kaplan chartered a plane and invited Rabinowitz and the carnivore specialist Luke Hunter (also of the WCS) to join them. Hunter had done extensive work with lions and cheetahs but had never seen a jaguar.[*] For Rabinowitz, it had been decades.

At Porto Jofre in the Pantanal, the three men—Kaplan, Rabinowitz, and Hunter—searched for jaguars. The pivotal moment came when their guide spotted a big cat fording the river. They followed at a safe distance. When the jaguar reached the shoreline, it shook and jumped up onto a small parapet. Then it turned and looked at the men. It was a big, powerful alpha male in the prime of his life, the essence of the Pantanal's wildness.

"It was orgasmic," Kaplan says of the sighting.

Back at the lodge, he asked the manager if there were any ranches for sale in the area.

The manager looked at him, clearly puzzled. "They're all for sale," he said.

A great flood in 1974 had killed 90 percent of the cattle in sections of the Pantanal, and most of the cattlemen who hadn't already sold out were willing to part with their ranches for dimes on the dollar.

"I want them then," Kaplan responded.

---

[*] In 2007, Penguin Random House South Africa published Hunter's seminal book *Cheetah*.

Wildlife biologist Raíssa Sepulvida tracking a male ocelot's movements using telemetry, Panthera Ranch, São Jofre, Pantanal, Brazil. (Sebastian Kennerknecht)

Female jaguar (Ague) after swimming with caiman kill across the Piquiri River. (Sebastian Kennerknecht)

Alan Rabinowitz with jaguar skull, Pantanal Brazil. (Panthera)

Alan Rabinowitz and George Schaller, Laos, mid-1990s. (Salisa "Tae" Rabinowitz, Panthera)

Jaguar and cubs, Corixo Negro, Encontro das Aguas State Park. (April Kelly)

Jaguar snarling at giant river otters. (Abigail Martin)

Alan Rabinowitz and Howard Quigley with a jaguar they captured and collared. (PANTHERA)

Mural, Bolivia. (JAMES CAMPBELL)

Abigail Martin piloting the boat on the Piquiri River. (JONAS STEINER)

Jaguar sleeping tree. (Joares

Uncommon black jaguar, Brazilian Cerrado. (Joares May)

Jaguar biologist Alliso Devlin tests a collar u telemetry. (Sebasti Kennerknecht)

Chievo, an adult male jaguar, Pantanal, Brazil. (Ezra Garfield)

Firefighters, Pantanal, Brazil. (Joares May)

Unidentified jaguar, Pantanal, Brazil. (Ezra Garfield)

Drone shot of Calakmul temple. (Taylor Turner)

Male jaguar (Juru) chasing black skimmers through shallow river water, Pantanal, Brazil. (Sebastian Kennerknecht)

Lesser anteater, Sa Miguelito Jaguar Conservation Ran (Duston Larsen

Alan Rabinowitz tracking tigers, Bhutan. (STEVE WINTER, PANTHERA)

Joares May, Fernando Tortato, and Raíssa Sepulvida tending to female jaguar Sofia. (WAI-MING WONG)

A six-month-old male jaguar cub getting a ride from swimming mother, Pantanal, Brazil. (SEBASTIAN KENNERKNECHT)

Champeon and his ranger team, Corcovado National Park, Costa Rica. (James Campbell)

Tom Kaplan and Alan Rabinowitz. (Luke Hunter, Panthera)

An unidentified male jaguar drinking at a river. (Ezra Garfield)

"Which ones?" the manager asked.

"All of them," Kaplan replied.

Going big was Kaplan's way. When he had been collecting Dutch art—seventeenth-century Dutch Golden Age paintings, including works by the Dutch masters Rembrandt and Vermeer, and buying on average a painting a week for five years—art leaders had been astounded. No one had ever done that before. He was the largest collector of centuries-old Dutch art in generations. Now he was doing something else that was unprecedented: mobilizing his zeal and his capital for big cat conservation. Ultimately, he was forced to settle for 165,000 acres of ranch land; if not for the byzantine Brazilian bureaucracy, he would have bought millions more.

As Kaplan and Rabinowitz worked to turn their portion of the Pantanal into a haven for jaguars, they discovered they were both attracted to the improbable. Each saw in the other something that compelled him. Rabinowitz made it possible for Kaplan to pursue his dream of big cat conservation. Meanwhile, for Rabinowitz, Kaplan was a "gift from heaven."

When Kaplan describes his blossoming friendship with Rabinowitz, he uses the Hebrew word *besheret*, which means "meant" or "destined to be." "We were brothers from another mother," he says. "We recognized that. We were both passionate, fearless, and forceful. But we were different, too."

Kaplan is fond of talking about what he calls the early days of Panthera. "There were many times after being on the river when we sat on plastic chairs, drinking caipirinhas or Venezuelan rum, smoking cigars, watching the day come to an end, and talking about jaguars and our dreams for Panthera. It was so exhilarating for both of us. We called it 'cat crack.' It was a huge endorphin rush."

Kaplan donated 25,000 acres of critical jaguar habitat along the riverfront to Panthera. As the project took hold, the number of jaguars grew. At the same time, the ecosystem was flourishing. By saving jaguars, he was saving other species that were integral to the environment—tapirs, giant anteaters, giant river otters, capybara, ocelots, pumas, maned wolves, and threatened species of birds. For

Panthera's field biologists, it was a dream come true. On Kaplan's ranches, they could study jaguars and other animals by actually watching them. Nowhere else in Latin America was that possible.*

Kaplan's vision, once considered unthinkable, was to create ranches where jaguars and cattle coexisted. Another conservation group had bought a ranch downriver. When the owners purged the ranch of its cattle, locals were incensed. They wanted to know who the gringos messing with their culture were. Kaplan was determined not to commit the same sin. He had, after all, done his PhD at Oxford on counterinsurgency, where the most basic lesson is that you have to win the hearts and minds of the people. "From a social point of view, the other ranch failed," Kaplan says. "It's called soft power. They overlooked that."

He laughs about it now, the implausibility of a vegetarian owning and running a cattle ranch in the Brazilian Pantanal, with an organization devoted to wild cats that was run by the world's foremost jaguar biologist—who had an allergy to cats. He made it known that although he would raise cattle, his property was a safe zone for jaguars. Under no circumstances would jaguar blood ever be shed. Eventually, Kaplan invited other ranch owners to take part in the reimagining.†

"People in the area took notice," he says. "Prior to us, the jaguar was viewed as a threat or at least a nuisance. But after we implemented our program, people saw that the jaguar was essential to achieving a better life for their kids."

---

* Kaplan also built small houses for the families of the cowboys and a bunkhouse for the other cowboys. Then he built a school and brought in a teacher. During the day, the children of the workers would get a free education, and at night the workers, most of whom were illiterate, would go to school. Thanks to a relationship with the dean of Mount Sinai Hospital in New York, he brought in residents, who administered free medical care.

† One of them was Mario Haberfeld, once a regular on the Formula 1 racing circuit. Haberfeld began the Oncafari Association at the Caiman Ecological Refuge in the southern Pantanal on the same ranch where Schaller, Quigley, Crawshaw, and Rabinowitz conducted early jaguar research. Like Kaplan, Haberfeld maintained the tradition of raising livestock, but he also introduced strict conservation principles and, eventually, ecotourism.

Slowly, even among the deeply traditional ranchers, the message spread, and Kaplan began dreaming of a loose confederation of ranches dedicated to the principles of big cat conservation. His ranches, Jofre Velho and, across the river, São Bento, formed the connective tissue along the Cuiabá River, which was bordered by a state park to the north, a national park to the west, a state park to the east, and a collection of private reserves to the south. They became the showcase for what many consider the "most ambitious carnivore conservation program in history." The man whom *Forbes* magazine dubbed "King of Cats" proved that an organization could take "rigorous science, backed by focused capital, to affect real change, to achieve measurable conservation outcomes."[2] Gradually the species flourished. Kaplan had hit upon what he calls the "secret sauce": no hunting, a plentiful food base, and room to roam.

Even those who doubted the sincerity of his intentions were surprised by the intensity—and the financial largesse—of his mission: "To harness the best minds, the best science, and the most capital into species-wide conservation." His goal for Panthera was to be "the Manhattan Project for cats."

If Alan Rabinowitz was Panthera's Robert Oppenheimer, then Luke Hunter (who would function as its president and chief conservation officer) and Rabinowitz's longtime friend Howard Quigley together played the role of Edward Teller. After his years in the Pantanal studying jaguars, Quigley launched a study of the Siberian tiger, *Panthera tigris altaica*, in the Russian Far East. There he tracked, snared, and collared tigers across the taiga. After leaving Russia, he went on to oversee the Teton Cougar Project, in the southern Yellowstone ecosystem, where he followed cougars on foot and on horseback throughout the mountainous West.

Another one of the best minds was George Schaller who, like Rabinowitz, Hunter, and Quigley, was working at the WCS. In his seventies by the time Kaplan approached him, Schaller's can-do days in the field, as a "feral biologist," might have been coming to an end, but he welcomed the prospect of consulting for Panthera (he also continued to work at the WCS) and participating in an organization that had the vision and financial muscle to accomplish its goals. Schaller would become Panthera's vice president. By that time, he viewed his work less as that of a biologist toiling in the field under harsh conditions than that of an "ecological missionary."[3]

A year before hiring Howard Quigley, Kaplan convinced Rafael Hoogesteijn to manage his ranches. In the 1980s Hoogesteijn had been part of a trio of groundbreaking biologists who put Venezuela on the map by initiating some of the earliest studies on jaguars. He was also a veterinarian, who for three decades had supervised two of the largest beef-cattle companies in the Venezuelan llano. There he was part of a team that experimented with measures designed to reduce cattle depredation by jaguars. It was the first of its kind in Latin America.

Although the evidence showed that livestock predation rates by large felids were vastly inflated, ranchers still claimed that jaguars (and pumas) had a significant impact on their cattle herds. Jaguars were often persecuted because of real or perceived threats to livestock.*

Under Hoogesteijn's supervision, Jofre Velho explored techniques that had never been tested before. One was vaccination. Most ranchers in the Pantanal didn't bother to vaccinate their cattle, exposing them to epidemic disease and parasites. Another innovation was night enclosures, with solar lights, well-lit corrals for pregnant females and calves, and eventually electric fences that packed a five-thousand-volt wallop. Hoogesteijn also experimented with collars, with reflective adhesive materials and bells, and with guard animals, especially water buffaloes. He brought in long-horned Pantaneiro cattle, which had been introduced to Brazil during the Spanish expeditions. Though smaller than the more common Nelore beef cattle, they were feisty enough to fend off jaguars. They were also durable and largely impervious to tick and worm infections. The most effective strategy, however, was probably the most basic one: making sure jaguars had plenty of prey to eat by outlawing the killing of their natural prey.

---

* That jaguars kill cattle was not in question, but in 1979 Schaller noted that cattle management in the Pantanal was so poor that only one cow in four or five successfully raised her calf to maturity.

While Hoogesteijn was struggling to change ranchers' hostile attitudes toward jaguars and to show that cattle and jaguars could coexist, Kaplan was using his considerable financial resources to fund the educations of doctoral students. Overlooking the Cuiabá River, watching waves of wood storks glide by, he and Rabinowitz sifted through candidates' applications and sometimes read them aloud. These generous "Kaplan grants" eventually funded more than sixty doctoral students. For Kaplan, it was an investment that made good sense. "If I was going to change the nature of cat conservation," he says, "I realized I would have to create a pipeline for talent, for future Alan Rabinowitzes, Howard Quigleys, and George Schallers."

In the meantime, an emerging ecotourism industry, based on a reimagining of jaguar conservation bolstered by the Panthera paradigm, took hold on the Cuiabá. Then Rafael Hoogesteijn and Fernando Tortato succeeded in putting a dollar amount on jaguar tourism. A groundbreaking study, the first of its kind in the Americas, it proved beyond the shadow of a doubt that the jaguar could be a financial asset instead of a liability.*

---

* The crux of the paper was that the approximate land-use revenue from jaguar tourism was three times higher than from traditional bovine cattle ranching.

CHAPTER 20

The Pantanal

IN THE SMALL DUSTY FRONTIER TOWN OF POCONÉ, THE "capital of the Pantanal," an archway, made of three large logs, announces to travelers that they are on the Transpantaneira Highway, greeting them in both Portuguese and English. Thereafter the pavement turns to red dirt, rock, and dust. There are exactly 122 bridges to cross. The road, raised six feet above the surrounding floodland, cuts a ninety-two-mile, north-south line as straight as a row of soybeans through the heart of the Pantanal, and it dead-ends at the sleepy backwater town of Porto Jofre, the Cuiabá River, and Panthera's Jofre Velho Ranch. In the wet season, when the Pantanal, the world's largest continental wetland, becomes an inland sea, the one-lane bridges, often nothing more than rickety planks built atop split timber log supports, make possible the passage over many of the ponds.

Brazil is larger than the contiguous United States by approximately 300,000 square miles, an area bigger than the state of Texas, and the Pantanal is wilderness on a Brazilian scale. At 42 million acres and covering almost sixty thousand square miles, the Pantanal, whose name derives from the Portuguese word for "swamp," is twenty times the size of the Everglades, more expansive even than the Okavango in Botswana.[1]

I arrived in the diluvial Pantanal at the tail end of the dry season, before the deluge and the swarms of mosquitoes. Our man behind the wheel, the Brazilian biologist Fernando Tortato, for the sake of speed, sometimes avoided the bridges, driving around them through the powdery remnants of what in a month would again be murky lagoons full of head-high water. Tortato explained to me that during the wet season, hundreds of billions of trees in the Amazon pull water up out of the ground and release trillions of gallons back into the sky as water vapor. Amazonian plumes rise from the canopy and collect twenty thousand feet overhead in giant migratory clouds. These "flying rivers," or *rios voadores*, as they are so poetically called, release their water as rain and transform the Pantanal's thirsty landscape into a water-logged wetland. Fed by swollen rivers of runoff from the Cerrado, an expansive savanna just south of the Brazilian Amazon, nearly the entire floodplain of the Pantanal fills with up to nine feet of turbid water and seeps north to south, across the Brazilian border into Bolivia and Paraguay. After inundating the Pantanal for nearly half the year, the water recedes from the maze of marshes, grasslands, wood pockets, and streams and drains back into the Paraguay River, leaving behind a verdant grassland flush with nutrients. Some say that the process, called the "flood pulse," imitates the behavior of a beating heart.

For the time being, the Pantanal's hydrology remains intact. Should newly unveiled plans be approved, however, another hundred dams will be built on the upper Paraguay River. According to Fernando Tortato, who co-authored an open letter decrying the expansion, the destructive dams would cut off the Pantanal's lifeblood, the flow of sediment and nutrient-laden water from the north. The entire ecosystem would be irreparably damaged.[2]

Tortato works as the manager of Panthera Brazil's conservation program and spends significant time at its 25,000-acre Jofre Velho Ranch, which sits on the edge of the Cuiabá River, a coffee-colored tributary of the sixteen-hundred-mile Paraguay, in the Brazilian state of Matto Grosso. The Portuguese *matto grosso* translates in English to "thick brushwood country." The state was once the largest in Brazil, before it was divided into Matto Grosso and Matto Grosso do Sul.

Joining Tortato and me on the long, cramped ride from Cuiabá to the Jofre Velho Ranch was Joares May, a Brazilian veterinarian of considerable fame who, while spending two hundred days per year in the field, had led capture and collaring efforts of jaguars, ocelots, panthers, tapirs, capybaras, and maned wolves all across the continent. May had been tutored by the South African Dr. Guy Balme, Panthera's director of conservation science and a renowned leopard biologist; and Dairen Simpson, a world-famous animal tracker and trapper who had captured and collared apex predators across the globe. May had trapped more than one hundred jaguars, including a big male in the western Pantanal that he named "Big George" in honor of George Schaller. In addition to his work as a wildlife vet, he was also a professor at a Brazilian university (Centro Universitario Univinte), a popular lecturer, and a local radio personality who does a lively weekly show on conservation and preservation.

The biologist Allison Devlin (now the director of Panthera's jaguar program), rode shotgun. In many ways, the engaging but unpretentious Devlin was one of Alan Rabinowitz's prized recruits. If he was Panthera's luminary, Devlin, and ambitious biologists like her across North and South America, were its future. Early on Rabinowitz recognized the need to recruit talented women like Devlin as well as homegrown biologists like Tortato, who would come to replace the old-white-male-savior icons who had dominated the field of wildlife biology for so long. Jane Alexander suggested that Rabinowitz's inspiration to encourage, and select, scientists from outside the normal recruitment pool derived from his experience as a marginalized stutterer. "He understood what it was to be an outsider and wanted to give people who did not fit the traditional mold a chance," she said. "But they had to be able to climb the mountain. Alan had high expectations. They had to be able to do the work."

Devlin had always had a predilection for animals, especially big cats. As a high school teenager, she got involved in an internship program at the Bronx Zoo. Because the zoo was also the WCS headquarters, she was able to attend talks that wild cat researchers often gave. As she came to know them, she expressed her interest in becoming a big cat biologist. They told her that if she was serious about following in their footsteps, she needed a hard science background. At Cornell, that's exactly what she decided to do. During her first year in the animal science program, she wrote a paper on jaguars. While working on that paper, she learned that Rabinowitz was scheduled to give a talk at the 92nd Street Y in New York City. Devlin's mom, who had once asked her why she couldn't be interested in butterflies instead of big cats, joined her for support.

After the talk, Devlin summoned the courage to introduce herself to Rabinowitz and confessed her nascent interest in jaguars. Impressed, Rabinowitz went out of his way to help, connecting her with two WCS researchers—Drs. Rebecca Foster and Bart Harmsen—in the Cockscomb Basin Wildlife Sanctuary in Belize. It was Devlin's big break, as it had been for Rabinowitz. For two summers she volunteered with Foster and Harmsen and eventually developed a master's project. For her master's degree at Columbia (and in collaboration with American Museum of Natural History and Cornell University), she was funded by a Kaplan scholarship to research captive and wild jaguars. She immersed herself in the program and even became certified as a scat detection dog handler. She used those skills to follow in Rabinowitz's footsteps, collecting almost one hundred jaguar scat samples in the Cockscomb Basin Wildlife Sanctuary that would have gone undetected if not for her black Labrador Shadow. By the time she concluded her program at Columbia, she had already developed a Ph.D. project at SUNY College of Environmental Science and Forestry at Syracuse. Now she was doing her postdoc at the University of Montana (funded by Duncan and Ellen McFarland), focusing on human-wildlife conflict in the jaguar range.

Because of Covid, it had been two years since Devlin returned to the Pantanal. In September 2021, the pandemic was on a downward trend in Brazil, but at its height, it had been especially bad. Hospitals in Rio de Janeiro were so overwhelmed with cases that they were on the verge of collapse.*

When Devlin explained how much she missed the Pantanal, she used the Portuguese *saudade*, a lyrical word that perfectly described the longing she felt for a place she had come to love. While doing her doctorate, she spent three months a year in the Acurizal Reserve and Pantanal National Park, where the Paraguay River and the Amolar Mountains to the west form the border between Brazil and Bolivia, and three months at the Panthera ranch. She said that when she saw the ranch and the Cuiabá River again, she would feel *matar as saudades*, "fulfilled." Then she laughed. A self-taught Portuguese speaker, she admitted that her accent was convoluted, half Pantaneiro and half city. The "Pantaneiro" dialect is to the urban São Paulo dialect what West Virginia hillbilly is to the English spoken in Greenwich, Connecticut. May, who had done an internship at the Smithsonian Institution, entertained us with an imitation of what he called "American redneck," saying that some of the people he had talked with in the United States sounded like twangy banjos.

Lying smack in the middle of the South America continent, the Pantanal was formed by the staggering convergence of more than one thousand rivers and streams spilling down from the Andes to the west and the high plateaus of the Cerrado to the east. For hundreds of years, cartographers had labeled the Pantanal *terra incognita*, and it developed a dark history of undoing the men who attempted to tame it. The legendary British adventurer Colonel Percy Fawcett, searching for the "lost city of gold," had entered the mazelike Pantanal in the early 1900s and barely reemerged. The writer Julian Duguid, who in *Green Hell* chronicled his 1920s expedition across Bolivia and parts of Brazil, added further to the Pantanal's sense of mystery. He described it as a "vast primeval sponge" that was largely "trackless" and unknowable to

---

* A congressional panel recommended bringing criminal charges against the president, Jair Bolsonaro, for his handling of the pandemic, including "crimes against humanity." Ultimately, no criminal charges were made.

all but its native inhabitants.³ In *Wild Things, Wild Places*, Jane Alexander called it "the Serengeti of South America."⁴

Nearly everything living in the sprawling Pantanal mirrors its physical size—the caimans, the catfish of the Cuiabá, the endangered giant otters (*ariranhas* in Portuguese), the giant anteaters, the jabiru storks that make their huge, treehouse-like nests along the river, the anacondas, and most important, the jaguars. The Pantanal grows jaguars like Iowa State grows offensive linemen: very big. In most areas of its range, the jaguar, or *onça* as it is called in Portuguese, had survived because of its adaptability, by becoming slyer, shyer, and smaller. But not in the Pantanal. The Pantanal ecosystem boasts Latin America's largest cats. One male collared by Panthera biologists in 2008 weighed in at 326 pounds, the size of a female African lion.

As we drove deeper into the Pantanal, Tortato handled the truck more carefully and encouraged me to keep my eyes peeled because drivers sometimes saw jaguars on the road, as well as brocket deer, tapirs, anacondas, crab-eating foxes that liked to dash in and out of the long grass, hardy Indian Nelore cattle that presented a danger to nighttime drivers as they bedded down wherever they pleased, and sometimes pugnacious Indian water buffaloes that refused to let vehicles pass.

Tortato avoided yet another bridge and spoke about the effects of industrial agriculture and habitat loss on the Pantanal and the Amazon. Easygoing and genial—what really stirred him were jaguars and soccer—when he talked about the ongoing destruction of Brazil's ecosystems, he spoke passionately. Ironically, what had become one of the main drivers of habitat loss in Brazil was the economic and political slugfest between the United States and China.

In 2018, during Donald Trump's presidency, the United States introduced $250 billion worth of tariffs on imported goods from China. Retaliating, the Chinese government slapped tariffs of 25 percent on $110 billion worth of U.S. goods, including soy. As a result, exports of U.S. soybeans to China, which amounted to almost 40 million tons in 2016, dropped precipitously (by 50 percent in 2018).

One of China's other major suppliers, Brazil (and Bolivia to a lesser extent), backed by promises from its president, Jair Bolsonaro, agreed to pick up the shortfall, which intensified land use changes that had become widespread well before Bolsonaro took office. For two decades, Brazil had been converting forest to vast fields of soybeans for use as animal feed and cooking oil. In 2016 alone Brazil lost 4,964 square miles of Amazon forest, an area larger than the state of Connecticut. Environmentalists protesting the wanton destruction risked their lives to decry, chronicle, and photograph the felling of the country's forests.*

Bolsonaro's trade deal with China involved a commitment unparalleled in Brazil's history. It put Brazil on course to increase its soy production by over 30 percent. That meant expanding its arable land by 32 million acres, a country-size chunk of ground as big as Greece. The only way for Brazil to meet its obligation was by weakening legal constructs that prevented large-scale deforestation, the erosion of Indigenous rights, and the loss of biodiversity. A team of scientists led by Richard Fuchs, a senior research fellow at the Institute of Meteorology and Climate Research at the Karlsruhe Institute of Technology in Germany, called on the United States and China to resolve their trade dispute in order to avoid an environmental "catastrophe" in Brazil.⁵

The state of Matto Grosso and its dusty towns, where cattle and soybeans reigned supreme and Bolsonaro was popular, had already been transformed by industrial agriculture. Once a backwater known for low-yield livestock ranching, it had grown rich, rolling in commodity cash, with the bright and glassy new city of Cuiabá, a five-hour drive from Porto Jofre. According to Tortato, that tumultuous development was unsustainable. The Pantanal, he said, was sitting on "a knife edge," its future determined by the financial viability of low-yield livestock ranching, which was increasingly threatened by high-yield cash crop farming. Sophisticated modeling, he added, also

---

* Between 2012 and 2016, at least 150 demonstrators were killed. Sandy, "Murder of Brazil Official."

revealed a troubling "arc of deforestation" advancing from the Pantanal north into the Amazon.

The booming city of Cuiabá, with its sudden skyscrapers spreading haphazardly from one corner of the city to the next, was founded in the early 1800s by gold-miner-explorers from São Paulo who mined placer gold and then imported sugarcane and cows. From the moment cattle were introduced, the Pantanal's cattlemen, who operated hundred-thousand-acre-plus ranches, or *fazendas*, that they ran like feudal empires, declared all-out war on the jaguar. Their hatred bordered on hysteria. Many ranches employed an *onçeiro*, a jaguar hunter. Being an *onçeiro* was like being a knight at King Arthur's table. *Onçeiros* acquired folk hero status by defending ranches from marauding cats that came in the middle of the night to wreak havoc on the vulnerable cattle. In *Through the Brazilian Wilderness*, published in 1914, former U.S. president Theodore Roosevelt, a conservationist and a lover of swashbuckling hunting tales, wrote dramatically of hunting jaguars on the huge properties of these cattle barons.*

While writing of his epic hunts, Roosevelt took the time to admire the Pantanal's wildness and to admonish the people of South America to take conscientious care of their lands and wildlife. "There is every reason," he wrote, "why the good people of South America should waken, as we of North America . . . are beginning to waken . . . to the duty of preserving from impoverishment and extinction the wild life which is an asset of such interest and value in our several lands."[6]

Despite these cautionary words, Roosevelt thoroughly embraced the mythology of the hunt. He described the jaguar as the "noblest beast" but also as a "brute" and an "enemy." He wrote of a "huge male, up in the branches of a great fig-tree," that his son Kermit had shot. "A bullet behind the shoulder," he continued, "from Kermit's 405 Winchester, brought him dead to the ground. . . . He had the big

---

* One photo shows Roosevelt kneeling. He wears a pith helmet and holds a rifle in his left hand, while he is cradling the head of a jaguar. With his right hand, he holds back its lips to reveal its large canines.

bones, the stout frame, and the heavy muscular build of a small lion; he was not lithe and slender and long like a cougar or leopard; the tail, as with all jaguars, was short, while the girth of the body was great; his coat was beautiful, with a satiny gloss."[7]

🐾

BY THE TIME WE REACHED BRIDGE 53, the sun was setting, its fall from the sky and its extravagant light accelerated by the approaching equinox. As it sank deeper, the late afternoon light dimmed, and the big savanna hawks ceased hunting for the day. Darkness came quickly. As we crossed the bridges, we saw large predatory eyes, constellations of them, glowing in the coal-black night.

"Caiman," Tortato told me before I could ask. Crocodiles.

In the 1980s yacare caimans were hunted by *coureiros*—leather men—who killed millions of them for their skins alone. Hunting animals to the brink of extinction is a very old story, but the *coureiros* were especially ruthless. To fuel the $100 million annual trade, they massacred the unsuspecting caimans, skinned them, and left the rest for birds and scavenging mammals or to rot in the blinding sun. The yacare caiman population plummeted. If not for the CITES Appendix II endangered species listing, a Brazilian government quasi-crackdown, and a 1992 global ban on the trade in wild crocodile skins, they would have disappeared. But the prohibition worked, and then, thanks to their ability to reproduce quickly, the species rebounded. Today nearly 10 million yacare caimans inhabit the Brazilian Pantanal, likely the largest single crocodile population on Earth. It's hard to love caimans, with their beady eyes, their stealth, and their fiendish hunting skills—they often swallow their food whole after drowning it—but their resurgence represents a remarkable victory for conservation and law enforcement.

"They're everywhere," I said, shining my flashlight along the edge of the lagoon where a ten-foot monster lay with its mouth agape. George Schaller had noticed the same thing when he first came to the Pantanal, describing it as a "couple of hundred eyes shining, like a city from an airplane."[8]

"Jaguar food," Tortato replied.*

As we walked back to the truck, and the first pulsing stars of the Southern Cross appeared, I glanced over my shoulder to see if I was being followed, wondering ludicrously if an overgrown caiman might regard me as part of the food chain. In actuality, caiman attacks on humans are rare. They prefer capybaras, the world's largest rodents, which look like overgrown woodchucks and can top out at 150 pounds.

By Bridge 90, we stopped again to stretch our legs and to tighten the ropes that were holding down the dust-lined tarp covering our supplies in the bed of the truck. When I finished cinching down one side, I turned off my headlight and was immediately enveloped by darkness. I scanned the horizon and didn't even see a far-off ranch light. Overhead, the arc of glowing galaxies and stars filled the sky. A wind out of the East brought the scent of fire. When my eyes adjusted, I saw tendrils of smoke rising in the distance.

As we loaded ourselves back into the truck for the final thirty bridges, I asked Allison Devlin why she had chosen to work on jaguar conflict. She said the conflict was about land use, and what, she asked, could be more important, especially in a place like the Pantanal? Although the Pantanal Conservation Area, a UNESCO World Heritage Site and Biosphere Reserve, consists of a cluster of four protected areas with a total area of 464,000 acres, 95 percent of its land is privately owned, with some 2,500 ranches running nearly 4 million head of cattle. These ranchers, in many ways, will determine the Pantanal's fate.†

In places like the Pantanal—where public land is in short supply, where land use, buoyed by escalating agricultural production, is changing, and where tremendous fires are becoming more common—

---

* Jaguars are especially fond of caimans, so much so that their disappearance in the 1980s may have depressed the big cat's reproductive capacity.

† Despite the alarming and growing threats to the Amazon, 80 percent of it is still intact, and jaguar populations there are holding steady—for the time being. In fact, just three national parks in the central Amazon have more jaguars than all of Central America.

jaguar populations are highly vulnerable. In short, if the jaguar is to have a fighting chance and if the connectivity of the corridor is to be maintained, the Pantanal and traditional landscapes like it—made up of a mosaic of public and private lands, where humans and wildlife share the same resources—are where the critical battles for the future of the species will be waged.

CHAPTER 21

Onça

I WOKE EARLY, BEFORE SUNRISE, TO THE INCESSANT screaming of Chaco chachalacas. Their call was something between the cackle of a happy hen that had just laid an egg and the chatter of a snow goose flying low over a field of picked corn. Their appearance also seemed to be a hybrid, a cross between an exceptionally large female quail and a roadrunner.

I walked down to the lagoon, where fish-eating bats flitted through the air and various shore birds, plovers, egrets, spoonbills, white ibis, and herons were hunting among the stoic and single-minded caimans, which they paid little attention to. I sat in the grass and watched downriver as the sun rose fire-red and light poured across the land. The Cuiabá River was big, wide, and silent. Away from the main channel, the land was dry and dusty, with areas of carpet grass and pockets of forest called *capoes* populated by various trees, including figs, elegant Genipa (*jenipapo* in Portuguese), Caesalpinia, acacia, ant trees, and statuesque dende palms. But I knew that this place would soon be transformed by water. In fact, there were small rises where I could see what appeared to be watermarks. Much of what I was looking at would soon be part of an expanding seasonal marsh.

Before breakfast, as the sun edged up from the earth, I made a brief reconnoiter of the ranch trails. Swallows sat on the power lines, and snorting toucans flew overhead—had I closed my eyes, I would have believed that pigs had indeed learned to fly. In the pasture, a horse rolled on its back, taking a dust bath before the day's heat set in. Farther back a few flowering Cambara and Piuva or ipê trees (*Vochysia* species) showed off their still-vibrant blossoms. As I walked, I scanned the sand and dirt for jaguar tracks, feeling the same sense of vulnerability that I'd felt in Costa Rica's Corcovado National Park. I wondered what I'd do if I actually saw a jaguar, or what the jaguar would do if it saw me. When Alfred Russel Wallace was in the Amazon, the Indigenous people warned him never to turn his back on a jaguar. A jaguar, they said, might bluff-charge a man, but it would attack only if his back was turned. Wallace, however, wasn't entirely convinced. "No doubt," he wrote, "a jaguar sometimes mistakes a man for legitimate prey."[1] In *Monster of God*, David Quammen points out that humans have always shared landscapes with dangerous alpha predators. Our *Homo sapiens* ancestors and flesh-eating monsters evolved together. Being attacked and eaten by a ferocious animal was a "grim reality" that human beings learned to live with. Quammen deadpans, "Among the earliest forms of human self-awareness was the awareness of being meat."[2]

The bulk of evidence, however, discounts the tales of the jaguar's threat to people. Costa Rican Eduardo Carrillo explained that although "a jaguar could eat any animal that crosses its path . . . there are few records . . . that jaguars have ever attacked people in the wild."[3] He tells a story of crawling through the jungle on his belly and coming face to face with a large male jaguar. The jaguar surely could have ended his life right there, but instead it watched him for half a minute and slunk back into the forest. Decades ago A. Starker Leopold arrived at the same conclusion, writing in *Wildlife of Mexico* that while he was doing his survey, he had never been able to authenticate a report of a Mexican jaguar becoming a man-eater. "Men undoubtedly have been killed by cornered or wounded jaguars," he added, "but unprovoked attacks are rare. In this respect the jaguar differs from its relatives in the Old World."[4]

Tigers, lions, and even leopards have a long, bloody history of man-eating. But not the jaguar. I had been told by a biologist friend that historically the jaguar kept to itself, preferring, almost always, to avoid confrontations with man. He said that some biologists believed jaguar mothers, when teaching their cubs to hunt, give them very specific "prey templates" that didn't include human beings.

One of the most reliable authorities in all the Americas on jaguar behavior lives for a portion of the year at the Panthera ranch. In 1992 Rafael Hoogesteijn, manager of the ranch, co-wrote one of the classic books on the jaguar, called, appropriately, *Jaguar*. Over the course of his career, he'd come across no fewer than forty jaguars while on foot. Hoogesteijn is a large, imposing man. Yet any one of the jaguars could have knocked him dead with one tremendous blow from its plate-size paw. He'd been growled at and bluff-charged but never attacked. "People," he said, "were never a staple food for jaguars. There are no cases of jaguars systematically killing and consuming human beings." Jaguars are smarter than tigers, lions, and leopards, he added: "They know enough to fear people" and cultivate an attitude of "nonaggression." Alan Rabinowitz had called them "reluctant warriors," which, he said, was a superior evolutionary strategy. "Humans who have lived or worked with jaguars all acknowledge the power, fierceness, and savagery of the animal and, at the same time, its nonaggressive nature toward humans. . . . One attempt to explain this phenomenon points out the fact that jaguars did not evolve alongside hominids."[5]

Hoogesteijn also explained that because of the *tigrilladas*, when jaguars were hunted to near extinction for their pelts, the bold jaguars died, while the shy and elusive ones survived. Today's jaguars are their offspring.

When I got back to the ranch, I walked to my room, following the stone steps to the barracks, a basic whitewashed building with a screened and covered porch that was used to house visiting scientists and, in non-Covid times, tourists. The boisterous Chaco chachalacas were screeching again, louder than ever, announcing the start of the day. Soon, I knew, the handful of *ribeirinho* (people who live along the river) kids who attended the small ranch school, named in honor of Alan Rabinowitz, would be arriving for their classes. Their cheerful teacher, Suliana, her husband Eduardo, a lead cowboy on the ranch, and their young daughter Eduarda lived in back of the main ranch house that, surrounded by barrel-chested mango trees, sat next to a building that housed the kitchen and cafeteria.

After breakfast, I went over to the main ranch house. Cats lounged on the cool bricks of the veranda. Inside, the common room sported big ceiling beams and a black-and-white-tiled floor that Hoogesteijn had rescued from the original ranch house. The room was dedicated to the jaguar. A map of the Jaguar Corridor hung prominently on one wall. Close by was a framed display of spearheads, *zagaias*, that were once used by jaguar hunters. At the back of the room was the requisite display of skulls, one of which was the horn of water buffalo with a jaguar tooth embedded in it.

In the front of the room, Joares May was giving Andrea Pisaro, a Colombian "jaguar woman" and a veterinarian by trade, a tutorial on how to handle a sedated jaguar. Pisaro had been with Panthera for only three months. Prior to that, she'd worked with chimps in Sierra Leone. She laughed that she'd gone from a societal animal, given to extreme moodiness that had been thoroughly studied, to one at the opposite end of the spectrum—a solitary animal about which wildlife biologists were still learning the basics. In Sierra Leone, she'd worked on human-chimp conflict, and in her native Colombia, she was concentrating on human-jaguar conflict. Much of that involved talking with chauvinistic ranchers, who dismissed her advice until they learned that she was a trained horsewoman and vet who could also tell them why their cattle might be getting sick. She said that she was happy to be back in Colombia, doing something positive for her country and protecting a species that represented Latin America's wild heritage.

As we were talking, Raíssa Sepulvida and Valeria Boron joined us and told us they were plotting an afternoon trip on the river. Sepulvida was Panthera's small cat biologist who was living at the ranch part time, while doing her Ph.D. on the movement of ocelots. Sepulvida, like Boron and Pisaro, was a testament to Panthera's efforts to promote women in field biology, to cultivate able scientists in their native countries, and to promote small cat research, which the NGO had overlooked in the past. Boron, an Italian, had recently come to Panthera from the World Wildlife Fund and was dividing her time between London and Colombia, while coordinating the implementation of jaguar conservation projects in Colombia, Suriname, and Bolivia. One of those projects entailed preliminary attempts to curb the rising trade in jaguar parts.

Once on the water, as our Pantaneiro boat captain expertly navigated the river and the floating masses of driftwood, I scanned the mud and sand flats and the riverbank, lush with tall barriguda trees (Bombacaceae) and woody vines, for any sign of movement. Jaguar numbers are difficult to measure, but at five or six jaguars per hundred square kilometers (or sixty-two square miles), the standard biologists use to measure density, the Pantanal has the highest concentration of jaguars in Latin America, perhaps two thousand in total, constituting what Howard Quigley called a "vital core population."

An hour after shoving off, and north of the Río Piquiri, we entered one of the side channels. The boat's bottom scraped on the sand, and soon we were part of a world more swamp than river, with floating carpets of water hyacinth and fire flag, and teeming with life. The first group of animals to greet us was a family of sleek and playful giant river otters that swam boisterously alongside the boat until the mood to dive overcame them. When they resurfaced, they did so, so far downstream that I wasn't sure they were the same animals. Newly arrived ospreys, black-collared hawks, and magnificent crested caracaras seemed to inhabit every tall tree along the channel. The largest tree, dead and stripped of its leaves, was the home of two jabiru storks, *tuiuiu* (pronounced *to-yu-yu*), including an attentive female, whose neck pouch looked to be packed with fish, and a fledgling that she was sheltering from the scorching afternoon sun. She stood motionless with her big wings outstretched. Egrets and a variety of stealthy herons hunted in the shallows. Jacanas, or Jesus birds, appeared to walk on water, as they moved among the water lilies and duckweed. Shore-hugging caimans, their mouths agape, hissed at us when we got too close. Capybaras lounged neck deep in the water. A huge green anaconda swam lazily under the branches of a tree that leaned into the channel. When the propeller of our boat motor hit a submerged log, a flock of horned screamers flew off, squawking maniacally.

About ten minutes in, we noticed two boats parked near the bank. A small group of Brazilian tourists were using their binoculars. One of the guides put two fingers up to his eyes and then pointed off into the distance. A tree stood about one hundred feet away, and I imagined that a jaguar sat under it, perhaps dining on a small caiman that it had dragged from the river, eating the brains and face first and then disemboweling the creature for its delicate organs.

As I waited for a glimpse of my first jaguar, it was as if time had been suspended. We might have sat there for five minutes or twenty-five. Had someone asked me, I wouldn't have known. I felt captivated by the imminence of the jaguar and would have stayed the entire day if not for Rafael Hoogesteijn, who broke the trance, saying it was time to look elsewhere.

We headed to the Cuiabá and back downriver. I continued to scrutinize the riverbanks. We saw one other boat, and our guide stopped to trade information. After that, he no longer slowed to investigate areas that he knew from past trips. The sun was preparing to sink, and we were bound for the ranch. Flocks of egrets returning to their roosts, big Muscovy ducks, and a noisy mass of bright green monk parakeets flew overhead. Our day on the river was over.

Days later, when Boron organized another trip, I jumped at the opportunity to be on the river again. As we passed the picturesque São Bento Ranch that Panthera had once owned, I was in full predator mode. I desperately wanted to see a big cat before I left the Pantanal.

I was sitting next to our guide, when just ten minutes upriver of São Bento, he slowed and then cut the motor. Then I heard him utter the word: *onça*. The sound barely left his lips. But he didn't need to say it. I had seen the big cat, fifty feet from our boat, on the riverbank, under the branches of an overhanging tree. Suddenly everyone aboard was silent. I was aware of my breath and tried to inhale and exhale as imperceptibly as I could.

The jaguar lay lazily in the sand, unconcerned by our presence or the rapid shutter *click* of cameras. I couldn't make out its entire form, but I could see its head, big and broad. It had a scar on its face. Its ears and tail twitched to ward off insects. As I was about to take a photo, the jaguar stood. It didn't jump to its feet. Nor did it get up wearily. It rose, as if every muscle in its body worked in perfect concert. Then it walked to the river and drank. Despite the heat, it didn't lap desperately at the water. It crouched and drank with a languid assurance. It held its tail high, but not once did it lift its head to look at us.

Then I heard Allison Devlin whisper, "It's Aju."

I thought Aju might return to his lair to lie down again, but instead he walked along the beach, near the edge of the water, even closer to our boat. It was then that I really saw him, his extraordinary power, his bull neck, his length, probably eight feet long from nose to tail, the musculature of his shoulders, the protruding belly that made him look as if he'd just consumed an entire caiman, the bulging testicles just under the tail, and the lethal canines of his half-open mouth. I had seen wolves, wolverines, grizzlies, polar bears, lions, elephants, and countless other wild creatures in their natural habitat, but no animal had ever captured my imagination like that jaguar. I glassed him with my binoculars, the dominant male of the Três Irmãos River, focusing on the pattern of his rosettes running along his rib cage. Seconds later he sprang easily onto a small ridge. He stood there for a moment watching the river, and then he was gone.

CHAPTER 22

The Pyrocene

ON AUGUST 5, 2019, ADECIO PIRAN, A LOCAL REPORTER and climate activist in Novo Progresso, a roughhewn town of thirty thousand in the Brazilian state of Pará, penned an article in the online *Folha do Progresso*. He warned of an upcoming *dia do fogo* (day of fire), when protesting local growers and cattle ranchers would set afire lands in the Amazon rainforest. *Queimadas*, illegal burnings to clear fields and forests, were common in Brazil, but the August 10 fires, Piran suggested, were inspired by rebellious people rallying around Jair Bolsonaro's Amazon policy. After Reuters picked up the story and it made worldwide headlines, Brazil's government accused Piran of being a purveyor of lies. Nabhan Garcia, Bolsonaro's special secretary of land affairs, called Piran's article mendacious and "unpatriotic" and demanded he retract it. Piran refused. When five days later Bolsonaro finally sent in state and federal police and two hundred soldiers, they encountered the smoking remains of what had once been a flourishing rainforest.[1]

Agamenon da Silva Menezes, leader of Novo Progresso's Rural Producers' Union, was defiant. "The Amazon is ours," Menezes said. "We will preserve the Amazon . . . according to our needs, not the needs of the world." Menezes knew his bold claim was heartily supported by the government, especially by Garcia, who had committed himself to modifying what he called the "embarrassment" of conservation areas and Indigenous lands expanded under previous governments.[2]

The Amazon forest, with its extraordinary biodiversity, holds one of the only jaguar subpopulations in the Americas that is not threatened with extinction. According to the wildlife ecologist Mathias Tobler, it contains "over 80% of the current occupied range of the jaguar," meaning that the future of the species is dependent not only on the Amazon biome but especially on its Indigenous Territories (ITs).[3] Together the land mass of the ITs occupies one-fifth of the Brazilian Amazon, close to half a million square miles, an expanse bigger than the Four Corners states of New Mexico, Colorado, Utah, and Arizona combined. Divided among several hundred reserves, some the size of European countries, they are occupied by more than two hundred culturally distinct Indigenous groups, totaling more than 800,000 people. Sixty of those Indigenous groups are considered extremely isolated, or uncontacted, and by Brazilian law entitled to special status.[4]

In June 2022 the biologist Joe Figel and a team of fellow researchers established the importance of Indigenous territories to jaguar conservation. In an attempt to identify areas of what they called "co-occurrence," they set out to systematically map the convergence of Jaguar Conservation Units and Jaguar Corridor lands with Indigenous Territories. To highlight his point, Figel demonstrated that some of the largest, most intact Jaguar Conservation Units in Colombia, Peru, Ecuador, Venezuela, and Brazil corresponded with designated ITs. In Venezuela and Ecuador, the overlap between JCUs and ITs exceeded 70 percent. In the Amazon, which includes the corridor's most expansive JCU, overlap was greater than 40 percent. Ultimately, what the group discovered was both encouraging and alarming: In Central and South America Indigenous groups had been exemplary stewards, reliably maintaining jaguar habitat and connectivity, but overall the ITs were "overlooked and undervalued" as conservation models.

Recently, another group of researchers corroborated Figel's findings. It found that Indigenous reserves are home to around 24,000 jaguars, or 63 percent of the total estimated number of jaguars in the

Brazilian Amazon. In fact, of the top-ten protected areas for big cats, eight of them were part of ITs. Sadly, the group also found that these territories fall within, or near, "the arc of deforestation"—the area where habitat degradation brought about by logging, ranching, mining, oil, road development, and fire accelerated dramatically during the Jair Bolsonaro administration.[5]

Destruction of the Amazon dates back to the 1960s, when Brazil's military dictatorship, which remained in charge of the country for twenty-one years (1964–85), promoted the area for colonization as part of its nation-building ideology. Brazil was modernizing, and its economy, bolstered by megaprojects, including bridges, dams, mines, and roads, was hailed as the "Brazilian Miracle." But no leader was as intent on developing the Amazon as a resource colony as Jair Bolsonaro.

Bolsonaro, who gave himself the provocative nickname "Captain Chainsaw," publicly sided with the loggers, miners, farmers, land grabbers, speculators, and agribusiness interests against environmentalists and Indigenous peoples (*isolados*) trying to protect the Amazon. While some of the colonists were poor, hardscrabble settlers, many others were representatives of land-grabbing criminal ventures. *The Washington Post*'s Terrence McCoy, who won a 2022 George Polk Award for environmental reporting, called the demolition of the Amazon "a crime story, at its core."[6] Indigenous peoples who dared to protest the destruction of their forests and invasions of public land sometimes disappeared. While defending invasions of the Indigenous Territories, Bolsonaro stated publicly, and with messianic zeal, how he admired the U.S. cavalry and its remorseless policy of eradicating the country's Native Americans.[7] He declared that he would not give "one more centimeter" of protected land to Indigenous peoples, vowed to take a "scythe to FUNAI" (the National Indian Foundation, the government's Indigenous rights body), and announced his intention to bring Brazil's Indigenous peoples, whom he called "cavemen," into the mainstream of Brazilian life.[8] Meanwhile his environment minister, Ricardo Salles, whom environmentalists dubbed "The Terminator," shamelessly suggested that the enormous fires sweeping across the country, creating columns of smoke rising to the stratosphere, were caused not by deforestation but by a perfect storm of dry weather, wind, and excessive heat.*

Although the Amazon basin is nearly the size of the contiguous United States, stretching over 1.4 billion acres and eight countries—Brazil, Colombia, Peru, Bolivia, Ecuador, Venezuela, Suriname, Guyana, and the territory of French Guiana—62 percent of it lies in the country of Brazil. Brazil is also responsible for the majority of the deforestation. So in 2019, just two weeks after the Day of Fire, when alarmed international leaders called for an end to the fires, they directed their comments to Jair Bolsonaro, whom many had taken to calling "Trump of the Tropics" for his belligerently antienvironmental rhetoric. Bolsonaro rebuffed them, accusing them of environmental imperialism and a form of "environmental psychosis." He even picked a fight with Amazon defender Leonardo DiCaprio.[9]

A year later Bolsonaro took notice when thirty-two international financial institutions, managing trillions of dollars in assets, threatened to divest if Brazil was unable to address the rampant deforestation of the Amazon. Just weeks later thirty-eight transnational companies sent a letter to Brazil's vice president Hamilton Mourão, whose job as president of the Amazon Council was to protect, defend, and sustainably develop Brazil's forests, urging him to address "environmental irregularities and crime in the Amazon and other Brazilian biomes." Bolsonaro got the message. Almost immediately he announced a 120-day ban on setting fires in the Amazon.[10]

As bad as the Amazon fires were, the Pantanal was hit even harder. In 2020, 8.1 million acres, or 27 percent of the Pantanal, burned. The fires were so dreadful that on February 2, 2022, World Wetlands Day, a collection of conservation groups sent an urgent letter to the secretariat of the Ramsar Convention on Wetlands, in Switzerland, requesting a full assessment of the damage and an accounting of the governmental neglect. They asked Ramsar to include the Pantanal on the Montreux Record, the official world list of endangered wetlands.

---

* In 2021 Salles resigned under pressure because of allegations he interfered in a police probe into illegal Amazon logging.

Their fear was that without a stronger national and international commitment to protecting the Pantanal, the fires would continue, creating a vicious yearly cycle that would devastate the landscape.

Like the Amazon, to which it is intricately connected, the Pantanal is actually shrinking. More deforestation in the Amazon ultimately means less rain to flood the great Pantanal basin. As it dries out, more and more fires will burn out of control, and the great inland marsh will lose its ability to function as a carbon sink.

Fernando Tortato, who has monitored the Pantanal's blazes for the federal government for the last two years, explained that fire, often ignited by lightning strikes, was endemic. Normally fires, circumscribed by the surrounding floodplain, exhausted the fuel load and extinguished themselves, but because of a disastrous drought and soaring temperatures that regularly reached 104 degrees Fahrenheit, with spikes surging to 115 degrees, ecosystem flammability was at an all-time high. Nearly 50 percent of the land burned in areas that for decades had been flooded and had no previous records of fire. Fires ignited underground, too, destroying a formerly fireproofed biomass of grasses, leaves, reeds, mosses, fibers, and trees. What's more, the smoke from forest fires introduced so many extra particles into the atmosphere that the vaporless clouds were unable to bring rain.

The headline of a *New York Times* article on October 13, 2020, could not have been more frightening: "The World's Largest Tropical Wetland Has Become an Inferno."[11] For two straight months, the Pantanal fires scorched the countryside, destroying crops. They jumped the borders of ranches and inundated surrounding Indigenous lands. Engulfed for weeks in smoke and falling ash, people were forced to fight them with anything they had. Ash infiltrated rivers, killing fish that people had come to depend on for food. People who breathed in particulate poisons for months complained of respiratory illnesses. Those who saw doctors learned that smoke could damage not just one's lungs but also one's kidneys and liver. Pregnant mothers worried about the effects on their unborn fetuses.

The main culprit, the one that cast an ominous cloud of smoke across the Pantanal, was climate change. For those who thought the fires that had transfigured the Pantanal were unexpected and unconnected to climate change, Fernando Tortato, like a prophet in the wilderness, had a sobering bit of news. What had happened in 2019 and 2020 in the Pantanal, one of the two largest remaining bog complexes in the world (the other is the Great Vasyugan Mire in Siberia), was not an exception; the bitter truth was that megafires were the new normal, climate change with a raging vengeance. We have entered what Stephen Pyne, fire historian and emeritus professor at Arizona State University, calls the pyrocene, when "fire creates the conditions for more fire."[12]

Models, in fact, show the Pantanal region alternating for decades between two ecosystem-damaging extremes: appalling heat and deep, persistent drought, followed by torrential rains, leaving the region lurching violently between two dramatic weather patterns. These intense systems represent, studies say, a "critical threat" to the health of the Pantanal, altering its biodiversity and its long-standing role in regulating water for all of South America and compromising its ability to soak up carbon for the continent—and for the world. Though wetlands like the Pantanal make up only 3 percent of the Earth's land area, they store a disproportionate amount of carbon, twice as much as all the trees on Earth. Yet in our efforts to address climate change, these wetlands have been largely overlooked.*

Fernando Tortato saw the desolation firsthand—miles and miles of charred land littered with the singed carcasses of animals that had been strangled by the smoke. In one dried-up mud hole, he saw ten tapirs that, once mired, hadn't had the strength to escape. One Brazilian journalist termed the unsuccessful flight of the animals the "death walk." Gustavo Figueirôa, a biologist working for SOS Pantanal, a conservation nonprofit, said, "We're watching the biodiversity of the Pantanal disappear into ash."[13]

What Tortato was able to save, at least in part, was Panthera's

---

* Tortato said 2023 was the worst year of all because fires burned from October through January, in the middle of the rainy season. That had never happened before. And the Paraguay River, because of the lack of rain, was at its lowest level in 124 years.

25,000-acre Jofre Velho Ranch, a critical link in the Jaguar Corridor Initiative. He joined locals, environmental agency workers, and members of the Brazilian military building firebreaks and dousing flames with the help of mobile thousand-liter water tanks and hoses that allowed them to pump water from the Cuiabá River and nearby lagoons. They'd hoped to keep the fire from crossing the Transpantaneira Highway, but when it jumped the road, they knew they were in trouble.

"I was looking at the NASA maps that show the locations of the fire, and I knew that they were spreading rapidly," said Tortato. "When the flames grew close to the fire break, we battled, fighting to control them. But at some point, we simply lost control—there were just too many places where fires were breaking out. At one point, in a very remote area—they seemed to almost jump from one side of a river to the other. That's the moment that we lost hope, almost. But the next day we woke up and started again."

Across the Pantanal, literally millions of animals died. A group of scientists measured the loss at 17 million vertebrates, including primates, birds, and reptiles. And vital habitat disappeared, too. Although the Jofre Velho Ranch was nearly surrounded by flames, ultimately Tortato and the small military team he was part of were able to save 20 percent, an area they knew would function as a vital refuge for the region's hard-hit wildlife. Jofre Velho had been lucky. Many other ranches were decimated. Panthera biologists estimated that the habitat of up to 500 jaguars had been destroyed in the conflagration.

CHAPTER 23

The Lucky Cats

THE PANTANAL FIRE WAS RED HOT, JUMPING LAGOONS and even rivers, burning madly in the tops of trees and smoldering below the surface of the Earth, in the prehistoric layers of steaming peat. Frantic birds caught fire and flew off in terror, spreading flames with their burning wings, before falling to the smoldering ground.

Ousado, a four-year-old, 175-pound male jaguar, had been trying to outpace the blaze for days, but his paws, scorched by third-degree burns, were beet red and rubbed raw, and he could hardly walk, much less run. Exhausted, Ousado, whose name means "bold" or "daring" in Portuguese, surrendered to the excruciating pain and bedded down on the banks of the Corixo Negro. A tributary of the Três Irmãos River, in the Encontro das Águas State Park, it had been engulfed by flames for months. Eighty-five percent of the park's 267,000 acres would ultimately be reduced to ashes, leaving behind a charred landscape and animal carcasses that rotted in the sun, while overhead, black vultures and caracaras circled in the rising thermals.

On September 11, 2020, a volunteer group of veterinarians from two nonprofit organizations, the Ampara Animal Institute and the Group for Animal Rescue in Disasters (GRAD), while patrolling the river for injured animals, spotted Ousado lying on his side and near death. He needed emergency treatment, but they would have to sedate him before transporting him to an aid station.

Jorge Salomão, an Ampara wildlife veterinarian, used his blowgun to shoot the first dart, but it bounced off Ousado's hindquarters. After consulting with another vet aboard the boat, he took his blowpipe and a handful of darts and left the safety of the river. Wading through the hip-high grass, he saw Ousado bedded down. When the big cat bared his teeth, Jorge froze. Forty minutes later, back at the boat, he formulated a plan with the others. He would return to the riverbank, and when he was situated, those on the boat would slap their paddles against the water, distracting the jaguar just long enough for him to get in position for another shot. The plan worked, and the second shot from Salomão's blowpipe found its mark. Once the sedative took effect, the men carried Ousado to the river and rushed him to the nearby Panthera ranch, Jofre Velho, which had been transformed into a wildlife triage station and firefighting command post.

Dehydrated, malnourished, and severely burned, Ousado was breathing painfully. While Salomão called for a helicopter, Carla Sassi, a veterinarian and coordinator of GRAD, struggled to stabilize the big cat. When the navy helicopter arrived, the vets loaded Ousado aboard. Salomão stayed by his side, accompanying him to the veterinary hospital of the Federal University of Mato Grosso in Cuiabá. From there the jaguar was rushed almost six hundred miles east, by van, to Corumbá de Goiás and the NEX (No Extinction) Institute, a Brazilian NGO, where vets treated the lesions on his paws and administered antibiotics.

Meanwhile, back at the Panthera ranch, Sassi worried about the other creatures endangered by the fires. She knew that if the jaguar, an animal capable of swimming and running, had been overtaken by the blaze, others would be considerably more vulnerable. She'd seen dehydrated tapirs, monkeys, coatis, and caimans, close to death, drag themselves to the food and water stations that her group had helped to set up along the length of the Transpantaneira Highway. Never had she expected to have to bring water to the soggy Pantanal.

Eventually, Ousado would undergo an intensive stem cell treatment, after which the big cat was transferred to a rehabilitation center to continue his recovery. It was a long shot, but everyone hoped that he might be returned to the wild one day.

Thanks to the herculean efforts of countless people, that auspicious day arrived even sooner than many had thought. On October 20, 2020, just over a month after Ousado was rescued, Fernando Tortato helped transport the cat back to the river where he'd first been found. Though fires still burned, they'd been largely controlled.

The team placed Ousado in a crate to which they attached a long rope. From the safety of a boat, they pulled the rope, lifting the crate's door. Ousado remained inside. Then he stood, took a few careful steps, and peeked his head out. He scanned the terrain, looking from side to side, and sniffed the air before he slunk off into an unburned portion of forest, with a radio collar around his neck. From the river, team members cheered and clapped. Ousado's collar had been programmed to collect data every hour for four hundred days until it automatically fell off.

Ousado's release coincided with the first rains of the wet season, and analysis of the data transmitted by his collar revealed that he had adapted quickly to his old home turf and seemed to be thriving. It was a happy ending to what was largely a tragic story.

"It was a victory," Tortato said. "It represented a new beginning for [Ousado] and symbolically for the Pantanal."

IN EARLY 2012 KYRA, the matriarch of Pantanal jaguars, gave birth to two two-pound cubs, Patricia and Jorge. They were born blind, deaf, and utterly helpless, but by the time Patricia and Jorge were five months old, and weaned, Kyra began introducing them to the world outside their den, taking them on short hunts and exploratory jaunts. Patricia was an astute observer and learned quickly. But Kyra was an overly vigilant mother, and it wasn't until nearly a year later that she allowed the cubs to wander independently. That's when the zoologist Abbie Martin and members of her Pantanal Jaguar

Identification Project, a not-for-profit organization that identifies and studies the northern Pantanal's jaguars, first became acquainted with the two cubs.

The Cuiabá River had always been a man's world, but Martin, who had a long history of flouting gender roles, was able to gain the acceptance of the proud and parochial Pantaneiro boatmen. Gradually, word got out that when she wasn't in the Pantanal, adeptly guiding a riverboat on the Cuiabá and studying jaguars, she lived for months in the U.S. Virgin Islands, where she was a hundred-ton master captain, in charge of a forty-five-foot sailing catamaran that took tourists through the Virgin Islands National Park and Coral Reef National Monument in St. John, U.S. Virgin Islands.

Shortly after Patricia and Jorge asserted their independence, they left their mother Kyra for good, striking out on their own, though they continued to hunt together and share their kills. By the end of September 2013, Martin noticed that Patricia had started to shun her brother. She speculated that the jaguars were becoming sexually mature and were prospecting for mates. She knew, too, that it was a crucial time for both. They would need to learn to live alone, a pivotal stage at which each would be at their most vulnerable.*

The day that Martin will always remember was September 28, 2013. As she wrote in her field notes, she and her team had been watching Patricia for four uneventful hours, while the big cat rested in the sand under the lip of the riverbank. Late that afternoon a boat upriver radioed Martin to let her know there was a jaguar moving in Patricia's direction. Martin fretted about the potential encounter. Apart from mother-cub reunions and mating dalliances, meetings between jaguars sometimes ended in aggressive displays of dominance or ferocious clashes. Martin had her binoculars fixed on Patricia. When the big cat caught the other jaguar's scent, she tensed.

As the approaching jaguar spotted her, it froze, ready for a confrontation. Patricia crouched, too, and laid her ears back, and the

---

* In the eight years Martin and her team have been documenting jaguars, they have identified 257 individual jaguars in their study area.

two jaguars eyed each other suspiciously. But then something unexpected happened. Neither jaguar leaped forward to fight. When Martin trained her binoculars on the intruding jaguar, a young male, she watched as it did the most astonishing thing. It rolled onto his back and exposed its belly. Focusing again on Patricia, Martin saw her pounce. But something about Patricia's action lacked aggression. When Patricia landed on the male jaguar, she didn't tear at him with her claws and teeth. In fact, it looked as if she were giving him an affectionate embrace. That's when it dawned on Martin. The intruder was none other than Jorge, Patricia's brother.

Sister and brother wrestled jubilantly, rejoicing in their reunion, and as the sun was setting, they strolled off into the trees together. It was the last time anyone from the project ever saw Patricia and Jorge together, but it was a reunion they would never forget. Jorge later moved outside the park boundaries and outside the purview of the jaguar ID project, while Patricia became a regular, settling in the middle of the park.

The following season Martin and her team observed Patricia mating with several of the park's dominant males. By the beginning of the 2015 season, they discovered her up the Três Irmãos River, far from her normal territory. But Patricia had a surprise for them: her first cub, a five-month-old male that the team named Capi. As they watched Patricia and Capi, they noticed that she was a very intense mother. She was exceedingly protective of Capi, as her mother Kyra had been of her. Sometimes Patricia would become so anxious, she would charge Martin's boat when it was anchored in the river. Although Martin was keenly interested in observing Patricia and Capi and learning about mother-cub hunting habits, she and her team were careful not to agitate Patricia, hoping that if they could earn her trust, she would allow them to watch and study her for many years.

The following year Patricia had another litter, two female cubs that the team named Jaju and Medrosa. This time, they noticed, Patricia seemed to be more relaxed and less watchful. Often she would allow her cubs to wander in the open and even to drink from the river with

Martin's boat no more than fifty feet away. Martin's strategy of not pressing Patricia was paying off.

Martin could see that the young cubs had very different personalities. Young Jaju, named for a local palm, was bolder than the timid Medrosa. Medrosa—the name means "fearful"—was leery and guarded and walked in Patricia's shadow and was usually the last to come out of hiding and the first to return to the cover of the riverbank after drinking. Slowly, however, as she became accustomed to Martin's boat, she grew less afraid, and by the end of the 2016 season, she had become fully habituated to the presence of people. For Martin and the Pantanal's guides, it was a very encouraging development. It showed that like lions, leopards, and cheetahs in southern Africa, jaguars could adapt to the presence of humans, which meant that jaguar viewing in the Pantanal could become an economic alternative for its subsistence farmers and fishermen.

In 2019 Patricia gave Martin another incredible insight into jaguar behavior. Martin and the other members of the project witnessed her mating with a new dominant male by the name of Juru. But Patricia had a three-month-old cub named Chula and was clearly lactating. According to all previous jaguar studies, a female jaguar could not enter estrus while she was nursing. Patricia had employed a tactic no one had ever seen before, deceit, hoping to trick the male jaguar Juru into thinking that that her cub Chula was his, in order to discourage infanticide. Often a dominant male will kill a female's cubs in order to mate with her. The following day Martin watched mother and cub swim the river together. But that was the last time she would ever see Chula.

Twenty-twenty brought destruction of a greater magnitude, as megafires ravaged the Pantanal. From her home in upstate New York, Martin was riveted to the frightening satellite imagery of the huge blazes. She watched powerlessly as they burned large portions of the Pantanal, and she doubted her beloved jaguars could escape the flames.

When she finally made it down to the Pantanal at the end of October, just before the wet season, she realized that many of the jaguars had somehow managed to survive. One of the first jaguars she saw was Patricia, the big cat she had come to care about so deeply, the one that had given her such extraordinary insights into big cat behavior. Patricia, she was overjoyed to report, looked strong and healthy.

CHAPTER 24

Return to the Pantanal

WHEN I REACHED THE TOWN OF PORT JOFRE LATE IN THE evening, I felt a curious sense of being back in a place that, inexplicably, had come to mean something to me. The Pantanal does not immediately inspire awe. From the north, there is only one road leading in. What one can see from that road, when not engulfed in dust, is more Pantanal, more savanna, swamp, and woods, with barely a hill or rise to catch one's attention. But when I arrived at the Jofre Velho Ranch, I knew what Allison Devlin had meant when she talked about *saudade*, her yearning for the Pantanal. I felt it, too.

The following morning I woke to the Chaco chacalacas and to the raucous calls of half a dozen hyacinth macaws. Decimated by the wildlife trade, hyacinth macaws are hard to find outside the Pantanal. But there they were squawking as if they hadn't a care in the world. A slight breeze carried a dose of wet season humidity, the fecund smell of the river, and the distinct scent of cattle. Gone was the sharp odor of smoke that on my previous trip had lingered in the air. In the distance, I heard the sound of thunder.

At the kitchen, I was greeted with hugs from exuberant staff members and uninhibited *bom dias* from the cowboys who were drinking *cafezinho*, their lavishly sweetened Brazilian coffee. Much of the talk was of a troublesome male jaguar that had been wandering the dirt roads of Porto Jofre. The jaguar, Juru, as I understood it, had killed a dog or two. It hadn't threatened anyone, but people were scared. Some talked of building a big fence around the town. Others suggested shooting the jaguar. But most urged patience, perhaps because many of the fifty or so families that called the somnolent river town home derived their income from jaguar tourism, and Juru was a jaguar that guides could count on when they had a boatful of hopeful clients. If not for jaguars like Juru, Porto Jofre would likely no longer exist as a destination, except perhaps for drug smugglers running cocaine between Bolivia and Brazil.

I had returned to the Pantanal and the Jofre Velho Ranch by invitation of Howard Quigley to participate in a Panthera capture and collaring project. Based on the information that Fernando Tortato had gathered from the motion-triggered camera traps, we were going to try to trap one or more of the resident females frequenting the immediate Cuiabá River corridor. The Panthera team was already getting valuable data from Ousado, who was fully healed. What they hoped to get from the females was location data that would yield home range estimates, behavioral patterns, and insights into the jaguar's "resource selection." Ultimately Panthera's biologists hoped to use that information to see which areas needed to be protected.

Howard Quigley was especially eager to have me participate in this effort. In his swashbuckling career, he had collared black bears, giant pandas in China (he was the first), Amur tigers in the Russian Far East (again, he was the first), and cougars across the mountainous West. Early in his career, he tracked jaguars in the Brazilian Pantanal by ultralight aircraft, nearly losing his life in a crash. In his ever-so-humble way, he dismissed the adventure and danger of his accomplishments, but he knew how exhilarating a capture campaign could be, especially for someone who had never taken part in one. He warned me that a campaign could be unpredictable. Although GPS collars had become irreplaceable tools for jaguar research since Alan Rabinowitz had used basic radio collars in Belize, capture projects were hardly effortless events. They had come a long way since Rabinowitz captured jaguars in primitive traps. But effective campaigns depended on the ability of the trapper to enter the jaguar's *Umwelt* and anticipate the jaguar's movements, exceptional teamwork, and a healthy dose of serendipity. They were high-energy, high-stress events because jaguars are smart and wary animals. Coaxing them into a snare, under the best circumstances, is difficult, even when a qualified wildlife veterinarian and trapper like Joares May, with a long history of successful captures, was in charge.

After breakfast, I joined May in the main room of the main house. He was conducting a gear check and had laid out his leghold snares as well as his medical kit, which he had personally designed for a Brazilian manufacturing company. The kit contained everything he would need to tend to a captured jaguar. He'd also set out a fly-fishing vest he wore to all captures that held needles and syringes, darts, blood tubes, rubber gloves, a blood glucose meter, and exactly two thermometers, two rolls of tape, two pens, and two pencils for recording important data. May laughed about his obsession with duplication but explained that he was abiding by an old Brazilian adage: "If you have one, you have nothing, and if you have two, maybe you have one."

Then he explained that order and discipline were the keys to a successful collaring. Keeping a darted jaguar safe was his top priority, and that required an unwavering protocol that he could perform practically with his eyes closed. He admitted that captures were adrenaline-charged events. Sometimes, he said, the hairs on his arms stood on end. But his job was to remove the emotion, to be unflappable, regimented, and always cool-headed.

May told a story about a night capture when he had snared a female in heat and was forced to perform his examination of the cat while her lusty male suitor hid thirty feet away in the bushes nearby.

In Brazil, a qualified veterinarian like May has to be present for all capture and collaring campaigns. For this one, Panthera needed the permission of the Brazilian government and needed, also, to be able to justify the campaign to the Institutional Animal Care and Use Committee. Sometimes it could take an entire year for the committee to review requests.

May described the exact steps he and the team of biologists would take if they were lucky enough to trap a jaguar. Depending upon the size of the animal, he would use 5 milligrams per kilogram of a drug called Telazol, which came in a powder form, to dart the jaguar. Telazol had a somewhat spotty track record, especially with older cats, but May preferred it to ketamine and had used it hundreds of times on animals. Its only drawback was that it took extra time to leave the body, dissipating in reverse order, the brain first, the muscles last, and sometimes it took hours for jaguars to regain their balance. May made it a practice to sit with a jaguar—at a safe distance—long enough to make sure the big cat adequately recovered from the narcotic.

May showed me the harness he used to lift jaguars when weighing them. Then he told me about a female named Noca that he had trapped in 2010, the first female jaguar ever trapped by the Panthera project in the Pantanal. It was a nerve-wracking capture. He was still a fledgling wildlife vet and had to perform under tremendous pressure in front of Panthera's hierarchy—Rabinowitz, Quigley, and Luke Hunter—as well as the journalist Bob Simon and a *60 Minutes* film crew. Noca weighed in at 160 pounds, a large female, and hoisting her required the strength of a number of men.

After the brief tutorial, I walked down to the boat launch for a day of jaguar watching before the collaring program officially kicked off. There I rendezvoused with our boat captain, the photographer Sebastian Kennerknecht, and Ezra Garfield, who had joined me on my trip to Bolivia. After getting skunked on jaguars in Bolivia, he had come by ferry, bus, train, and plane from San Juan Island, off the coast of Washington, to see a jaguar in the Pantanal. As a zealous birder, he was also hoping that the Pantanal would help him achieve his goal of seeing one thousand bird species by the end of the year. When I met him at the boat, he was all smiles, having just spotted a vermilion flycatcher sitting on a fencepost. The peripatetic Kennerknecht had just come to the Pantanal from India, where he had been photographing the elusive snow leopard.

Once on the Cuiabá, our guide's radio crackled. Minutes later we were speeding upriver toward the Três Irmãos against a strong current. When we arrived, a guide whispered that somewhere up on the elevated riverbank lay a jaguar.

Ezra saw it first, a male resting in the shade of a tree. We watched him with unbroken concentration for a good fifteen minutes, as he scratched his face, moved his head from one paw to the next, and flicked his tail. He did little else until he yawned three times.

"Here it comes," Kennerknecht said, alerting us.

The jaguar stood, arched its back, and yawned again, its mouth agape. From the boat, even without binoculars, we could see his tremendous size (later we'd identify him as a powerful male known as Chievo) and his bull-like head. Soon I saw people in the other boats lift their cameras. Then I heard the sound of dozens of shutters speedily clicking, breaking the wonderstruck silence. But it was Kennerknecht who got the perfect series of photos—the jaguar's wrinkled face and enormous canines—because he had predicted it. He had photographed big cats all over the world and said that three yawns almost always meant that a cat was preparing to move.

I wasn't sure what to make of the scene, two dozen ravenous tourists trying to capture the ideal shot. But in many ways, for jaguar tourism and conservation in the Pantanal, the clicking shutters represented the sound of success.

Because of Covid, I hadn't seen the conservation model in action the previous September. In fact, the river had been largely empty of tourists. But now I could experience what the Pantanal had become, Brazil's Kruger National Park, where lion, leopard, and cheetah sightings are virtually guaranteed.

The jaguar's recovery in the Pantanal is a remarkable conservation story. For decades, skin traders had killed them, trophy hunters hunted them, and stockmen shot them on sight. One Pantanal resident had even amassed a cattle kingdom as a jaguar killer. For every jaguar he eliminated, local ranchers paid him in cattle, two cows for every cat. But today, with the highest jaguar densities in the entire range, the big cats, and not cattle, pay the bills. Eradicating them no longer makes sound economic sense. Cattlemen, once considered the jaguar's feared enemy, are now sometimes among their most ardent protectors.

According to Jamil Rodrigues da Costa, a third-generation cattle farmer and the owner of the Porto Jofre Hotel, an ecolodge in the Pantanal, "People now realize that if the jaguar dies, so does the rancher."

Early on Alan Rabinowitz understood the jaguar's worth. He knew what George Schaller wrote in *A Naturalist and Other Beasts*, that "conservation problems are social and economic, not scientific," that human-apex-predator coexistence had to be predicated upon a mutually beneficial arrangement.[1] His hope was that he could help rebuild the Jaguar Cultural Corridor and resurrect a semblance of the respectful, man-animal relationship that had existed between the jaguar and the early peoples of Central and South America for millennia, while at the same time creating a model for environmental tourism by showing people that jaguars in the modern world could also mean jobs. He knew that people needed a material incentive for accepting jaguars. What was good for big cats needed to work, too, for the people living in close proximity to them, whose daily lives were enmeshed with them. In *Life in the Valley of Death: The Fight to Save Tigers in a Land of Guns, Gold, and Greed*, he wrote, "If any conservation effort is to be sustainable, the people most affected must view themselves as beneficiaries, not its victims. . . . Positive improvements to people's lives should be linked to the protected area or to the other conservation efforts that affect them."[*][2]

It was this model, a tapestry, that Covid had threatened to unravel. Jaguar populations had not yet suffered, but if another pandemic hit, and the economy didn't bounce back, what then? Desperate people would resort to desperate measures. Jaguar skins, teeth, and paste were worth hundreds of dollars on the illegal wildlife market, and opportunistic traffickers, operating across the range, who had grown bold during Covid, would grow even bolder.

On our way back, not more than twenty minutes upriver of the

---

* Quigley and Crawshaw called for the government to promote jaguar conservation by encouraging landowners' "interest in preserving parcels of their farms through fiscal incentives and partial tax exemptions . . . in addition to . . . idealism." Quigley and Crawshaw, "Conservation Plan." See also Franco, Drummond, and Nora, "History of Science," 52.

ranch, we saw two boats holding in the current near the riverbank. I knew now to look for the long-lensed cameras, which meant that a jaguar was nearby. As we pulled up alongside the boats and slipped our anchor quietly into the water, we saw why: a cat that Ezra would later identify as Patricia lay in the grass along the river. She was fully exposed, and we could see her beautiful markings. We watched her for about five minutes, and then she got up and walked along the riverbank, stopping every ten feet to gaze down at the water. Kennerknecht said he could see her belly was empty. She was hunting—when not resting, jaguars are almost always hunting—quietly and methodically searching the mudflats along the river for an unsuspecting caiman or capybara. Kennerknecht explained that jaguars hunt very deliberately. They never wander. They search for prey, careful, always, of not expending too much energy. Even in a prey-rich place like the Pantanal, jaguars lead a feast-or-famine existence.

Patricia walked in the open for a few more minutes, nimbly hopped a small branch blocking her trail, then disappeared in a tangle of vines and dense gallery forest. That's when the other boats moved on. But we stayed.

Not long after the boats sped away, Patricia emerged from the vines and slipped and slid her way down the steep riverbank. When she reached a small spit of sand, she stopped and looked at us, regarding us impartially. We watched in absolute silence. Then she looked upstream and down and eased herself into the river, swimming with the current as adeptly and fluidly as a giant river otter. I was surprised by her familiarity with the water, her deliquescence. Later that day, Raíssa Sepulvida, the ranch's small cat biologist, would say that jaguars are one of the most "water-loving" of all the wild cats. In the stifling Pantanal dry season, they will swim just for relief from the heat. They are also powerful, long-distance swimmers and can ford a mile-wide river channel while carrying a kill in their powerful jaws.

After swimming for fifty yards, she emerged from the water and shook each front paw, intently and comically, first the left, then the right. Then something captured her attention. Suddenly she was focused, laser-like. She turned back the flaps of her ears, each ear acting independently of the other, and waited perfectly still for a minute before climbing the high bank in two bounds. There she stopped again, listened for a few seconds, and stepped into the forest.

That evening, as we related our story of encountering Patricia, we heard the good news. Soon the trapping campaign would begin.

CHAPTER 25

Touching Wild

EARLY THE NEXT DAY I JOINED MAY IN THE FIELD TO help him reposition the jaguar traps, which had been washed out by three successive days of rain that were accompanied by an aberrant cold front out of Antarctica that tracked through a corridor between the Andes and Brazil's central plain.

With some modifications, May used a special low-impact snare designed by wildlife trapper Darien Simpson. The first thing he did was to position a small metal plate, into which he had drilled four holes, in the sand. I helped him drive two-and-a-half-foot-long pieces of rebar through each hole and into the ground. This represented the snare's anchor and would prevent a jaguar from dislodging the trap and dragging it off and injuring itself. Next, May attached a series of swivels to the plate, which were separated by a spring. The swivels and spring allowed a free rotation of the cable snare in case the animal resisted. To that, he added a three-foot-long thrower, equipped with an arm, a base, and a trigger. Last came the steel cable snare, which he laid on top of a thick sponge he'd embedded in the ground. The sponge functioned as a pressure plate. The idea was that when a jaguar stepped inside the loop, its weight would depress the sponge, which in turn would trigger the release lever and throwing arm and activate the tightening mechanism on the cable. May sifted dirt over the sponge and the cable to hide them as best he could, explaining that jaguars were gifted with lightning-quick reflexes and often were able to extract their foot before the snare even tightened.

Guiding a cautious and hyperalert jaguar into a leghold snare that measured one foot long by one foot across seemed an impossible task. To accomplish that, May used sticks, branches, and rocks to fashion a path that led directly to the snare. Then he brushed it clear of leaves, explaining that jaguars liked to walk silently. Getting the width of the trail just right was also essential. If the trail was too narrow, a jaguar would abandon it; too wide, and it would grow suspicious. Next, we cleared the "catch circle" of rocks, roots, logs, and anything else that could potentially harm a big cat as it struggled against the cable. When I asked May whether a jaguar would be scared off by our scent, he said that as small as the human population of the Pantanal was, it was still a human-dominated landscape, and jaguars had grown accustomed to the presence of people.

The last thing we did was to attach a transmitter to the trap. For that, Fernando Tortato had designed an amazingly simple but effective system. About fifteen feet from the snare, May laid down a tall bamboo pole that I had delimbed with a machete. At the top of that pole, he fastened a VHS transmitter, and to the bottom of that transmitter, he attached a magnet to which he'd tied a piece of fishing line. That line, once we stood the pole up and fixed it in the dirt, ran down the length of the pole, passed through a guide at the bottom, and was fastened to the snare. The idea was that when a jaguar stepped into the snare, the tension would pull the fishing line, which would dislodge the magnet from the transmitter and activate it, signaling that something had tripped the trap. Each of our six snares sites would have its own frequency, and once the capture campaign officially started, we would remotely check each site every hour, twenty-four hours a day. Before leaving, May placed a board over the top of the sponge, disabling the snare.

We had spent the better portion of the day preparing the traps, and when we returned to the ranch, we learned that Devlin had arrived.

After dinner, we all assembled in the main room of the main house, where Devlin and May prepared us for the campaign. Devlin stressed that captures were serious and sacred events during which we would hold the jaguar's life in our hands. Echoing May, she said that for a wildlife biologist, there was no greater privilege or responsibility. She also instructed us all to sleep in our clothes, so that if we got the signal that a trap had been tripped, we could mobilize quickly. Then she passed around a card with six frequencies, corresponding to the six snares that May and I had set earlier. All six frequencies needed to be monitored every hour, on the hour.

In the yard, under a nearly full moon, Devlin, who was holding a large metal antenna, showed us how we would monitor the traps. She held the antenna level and parallel to the ground and instructed us on how to check the receiver, just one click of the dial for every trap site. We listened as she clicked through each frequency. The signal for an empty snare would beat sixty times a minute. For a tripped snare, indicating a capture, it would beat faster, once every half second.

Back at our room, Wai-Ming Wong, the director of Panthera's small cat program, Ezra, and I carefully organized our gear. I laid out a headlamp, a mask (because of the possibility of a disease spillover—jaguars have receptors in their lungs to which Covid can attach itself), my camera, a rain jacket, an energy bar (Wong would bring the Haribo gummy bears), and an extra sweater because the nights had grown chilly.*

My slot, one of the graveyard shifts, was the three a.m. check, and I barely dozed that night. I was awake when Ezra went to check at eleven p.m. and when Wong went out at one a.m. When it was finally my turn, I walked into a yard so brightly lit by the moon, it looked like dawn.

I turned on the receiver and held the antenna. My breath came

---

* "Raised and educated in the U.K., Dr. Wong had studied sun bears and tigers in western Sumatra for his Ph.D. and, after joining Panthera, managed the tiger program for three years. Recently, he led the creation of Panthera's small cats program, focusing on the world's thirty-three small cat species. He now serves as Panthera's director of small cat conservation.

quickly, and I could feel my heart beating. What would it be like, I wondered, to be the one, to be the person to hear the 120 beats and then to alert the team that we had a capture? I turned on the receiver, turned the dial to the first frequency, and waited twenty seconds to be sure I knew what I was hearing. Nothing. Then I turned to the second, third, fourth, and fifth, each time expecting to hear a rapid pulse. When I arrived at the last frequency, my hope was tinged with fear. Would I recognize the signal if I heard it? I held my breath and turned the dial ever so slowly, half-believing that it would be the one. But all I heard was the slow, steady *tick tick tick*.

In the dining area that morning, everyone looked tired. Devlin announced that she, May, and Tortato had made the decision to keep the traps open all day. Under normal circumstances, especially in the Pantanal, where heat is almost always a factor, they would usually open the traps at five p.m. as the evening air started to cool and shut them down by seven a.m. But a front was moving through, and temperatures would be much colder than usual. That was both good news and bad. It meant that jaguars would roam and hunt less than usual, but it also meant that we had a twenty-four-hour capture window.

For the next five days, no one strayed far from the ranch, and no one had anything resembling a good night of sleep. And as the remaining days of the jaguar campaign dwindled, everyone grew a bit moody. Even Tortato's sunny disposition seemed to waver.

On our second-to-last day, we had a stroke of luck. Tortato told us over breakfast that the cowboys had discovered a dead cow in one of the pastures and that he, Devlin, May, and Sepulvida had decided to employ it as bait. Based on Sepulvida's camera trap photos, the cowboys would use the tractor and the bucket to lift and transport the cow to one of the trails that the jaguars appeared to be using regularly.

That night we all went to bed full of hope. Every hour I waited for the signal—a rap on my door that meant that we'd captured a jaguar. But the rap never came, and by the time one of the cooks sounded the gong for breakfast, I was already in the kitchen with May, who for the first time looked uncertain. May was a man who usually got his cat, and he was coming off a successful campaign at the Caiman Ranch in the southern Pantanal.

After breakfast, with May as my translator, I spent the morning talking with the ranch hands. One of them was an elderly man named Dito. In the 1970s, Dito had worked on a road crew helping to build the Transpantaneira Highway. He had been a machinery operator, though much of the work was done with horses and oxen and by brigades of men, under a blazing sun, using picks and shovels. The dream, Dito said, had been to build a road all the way to Corumbá on the Paraguay River in southwestern Brazil, hundreds of miles away, in the hope that it would bring access to markets and development. But the road, which never made it farther than the Porto Jofre on the Cuiabá River, fell far short of those dreams.

At the time, many millions of cattle grazed in Pantanal pastures, and as a boy, Dito took part in drives, when cowboys (*peão* in Portuguese), eating wild white-lipped peccary pigs as they traveled, drove cattle all the way to Corumbá. This was during the *tigrilladas*, so they rarely worried about jaguars. According to Dito, wild animals, in general, were harder to see then. Giant river otters, which were also hunted ruthlessly, because they were prized in the European fur markets, were so scarce, they were considered mythical creatures. Much of the Pantanal was grassland, as far as the eye could see, broken up only by occasional hammocks of trees, which the cattle used as refuge from the sun. After the torrential flood of 1974, when the landscape changed from grassland to swamp and forest, jaguar numbers grew, and people slowly began to recognize that the big cat presented a new economic opportunity. They became guides for eco-tourists, and rural economies began to recover. Although Dito was grateful for the financial opportunities the jaguar presented, he missed the old days when steamships, carrying salted fish and cured meat, plied the Cuiabá, and grassland stretched to the horizon.

By dinnertime, the mood had grown serious. There was little banter or laughter coming from the dining area. Everyone was painfully aware that it was our last capture night. We had been reduced to hoping that the stinking corpse of a cow would lure in a hungry female jaguar. When we left dinner for our rooms, Devlin, ever the optimist, reminded us that although we'd all been disappointed, our luck could still change. She'd seen it happen many times.

For the last time, I did a gear check before going to bed. At 10:33 p.m. Sepulvida rapped on our door. But no one stirred. It wasn't until she knocked a third time that I sleepily came to the door. Then I heard Devlin say in a commanding voice from the porch, "Let's go!" The capture was on!

Ten minutes later we were in the bed of the truck. When we arrived at the trap site, Tortato parked about fifty yards away. The jaguar's shining eyes pierced the night. It was moving back and forth, pacing quickly. May and Tortato approached it with the utmost caution, the light from their flashlight illuminating the big cat, and when they did, it ceased fighting the snare and lay down.

When May and Tortato returned, May informed us that we'd caught a female. I watched him uncase his rifle, a Danish Dan-Inject, Model JM darting rifle to which he and his father, who had been a captain and a marksman in the Brazilian army, had mounted a scope. Using a CO_2 air canister, the rifle would propel a dart, loaded with 5 milligrams per kilogram of Telazol. May and Tortato again approached the jaguar. She was about twenty feet away and still lying down when May, with perfect aim, darted her in the hindquarters.

May and Tortato returned and laid down the tailgate of the truck. May pulled out a form from the pocket of his vest, and while he waited for the drug to take effect, he filled out the date, time, and drug dosage. Then he checked his watch. Ten minutes later he and Tortato approached the jaguar. Confident the drug had done its job, they came back, and May explained what would happen next. He and Tortato would drive the truck to the capture site and load the inert jaguar onto the tailgate. May was concerned that because of the cow carcass, bulging with foul-smelling gases, another jaguar might be lurking in the nearby forest.

I watched Tortato and May lift the jaguar onto the tailgate. A small female, they handled her easily. Then as Tortato slowly backed up the truck, May walked behind with his hand steadying the jaguar.

In the light of the moon, I saw the cat, small, sleek, and beautiful, her head like a helmet and paws like winter mittens. Her coat was a mixture of color, polished gold along her back and torso, and cream on the underside of her belly and throat. Her rosettes were ornate, like flecks of coal, surrounded by dark, animated, imperfect lines that defined their shape. I smelled her musky scent.

"It's Sofia," Devlin said.

At 11:03 p.m. we turned on our headlamps, and the team began its work in a way that can only be described as reverent. Devlin gently wrapped a tape measure around Sofia's neck and realized that the female cat was small enough that Devlin would need to trim the collar for it to fit. The last thing she wanted was a collar that hung loosely from Sofia's neck and impeded her ability to hunt, or one that might fall off yielding none of the information she hoped to gather. Meanwhile, using another measuring tape, Sepulvida recorded Sofia's overall length, her girth, her height from ground to shoulder and ground to hip, the length of her tail, the circumference of her skull, and the size of her front paw. May dabbed ophthalmic gel onto Sofia's eyes to prevent her lenses from drying out. He explained that once tranquilized, jaguars lost their ability to blink. Then for further protection, he delicately covered her eyes with a bandanna to protect them. May said that he had been using the same good luck bandanna, his wife's, since his first jaguar capture in 2008.

To prevent infection, May applied antibiotic cream to the dart's entry point and onto the lower portion of Sofia's leg where the cable had pressed against it. Next, he clipped an oximeter to Sofia's tongue to monitor her heart rate and oxygen level. Using a stethoscope, he listened to her lungs. May's hands moved rapidly, but his voice, when he gave instructions, in both Portuguese and English, was calm and confident. Next, he opened Sofia's mouth to check the movement of her tongue and the tension of her jaw muscles and explained that if at any point he felt Sofia was becoming uncomfortable, he would administer a small dose of a muscle relaxant.

Then May did a urine draw and a blood draw, using Sofia's back leg, which he immediately tested for the distemper virus, a highly contagious disease that, among other things, can cause big cats to lose their fear of man. He pulled out a bit of Sofia's hair for DNA studies; the follicles would provide the team with a range of information, including the jaguar's eating habits, as well as a heavy metal reading. Because of a resurgence of illegal gold mining outside Poconé and the use of mercury to wash away the sand, biologists had grown worried about high levels in northern Pantanal jaguars.

At 11:26 Devlin announced that the collar, which included a small antenna, was set to send readings, including GPS points, for almost two years. The collar would provide a plethora of information about where she rested and slept, how much she moved on a given day, and when and how often she was eating.

Next, May wrapped a cuff around Sofia's tail and noted that her blood pressure was stable. By employing an instrument and app designed by a Brazilian company called INpulse, he got an electrocardiogram reading of Sofia's heart rhythm and electrical activity on his phone. Using a rectal thermometer, he measured her body temperature. Last of all, he checked the color and size of her teats and the condition and color of her vulva. When he announced that Sofia was in heat, Devlin took a moment to sweep the beam of her flashlight through the forest, looking for the gleaming eyes of a hidden male. All the while, frogs and insects trilled and mosquitoes swarmed around us.

While Tortato pulled ticks from Sofia's neck area and placed them in a small bag to be tested later for blood parasites, Sepulvida pulled back her lip to reveal a canine half as long as my index finger. That's when May announced that the examination was coming to a close. He and everyone else gave thanks to Sofia for her sacrifice.

The last thing they needed to do was to weigh Sofia. May and Tortato worked to fit her front legs into a harness. When they were certain she was secure, they hoisted her from the tailgate. She was small and weighed in at just over 123 pounds, perhaps half the weight of a full-grown Pantanal male. When they lowered Sofia back onto the tailgate, May announced that we could touch her. Sepulvida handed out rubber gloves and made sure we were all properly masked.

I touched Sofia first on the bulging muscle mass of her back leg. Then I ran my fingers along the length of her powerful tail. I rested my hand on her vulnerable belly, just underneath her ribcage, and felt the steadiness of her breath. Finally I stroked her smooth, luminous fur. My hand shook, and my heart fluttered. Among all the preparation, I had never prepared myself for the sheer rawness of emotion. The only similar experience I'd ever had was when my father died at home in bed, and my mom, brother, sisters, and I and all our kids sat next to him and touched his warm body. At the time, I remember thinking that I would never again be in such close proximity to the elemental(ness) of life. Then I came to the Pantanal and touched wildness.

CHAPTER 26

The Magic Cats

NINETEEN NINETY-SIX WAS A MOMENTOUS YEAR IN THE desert Southwest, when longtime Arizona mountain lion hunters spotted two jaguars: one in the Peloncillo Mountains in the New Mexico Bootheel, and the other in southern Arizona's remote Baboquivaris. The hunters had spent the better part of their lives wandering these ragged mountains. But the last thing they expected to encounter was a jaguar. The jaguar had been all but erased from the United States. More than thirty years earlier the last known female jaguar in the country had been shot and killed. In 1986 a rifle bullet toppled the last known male in the Dos Cabezas of southeastern Arizona.

At first Warner Glenn, a raw-boned fourth-generation rancher straight out of central casting, with clear blue eyes and a face coursed by deep lines, thought he was tracking a large mountain lion with his Walker hounds, his daughter Kelly, their wrangler, and a hunting client. After his strike dog caught the scent, Glenn was out in front, riding his sure-footed mule and following his baying hounds. Glenn struggled to keep within hailing distance. Eventually, the big cat escaped the dogs by scaling a high bluff. When Glenn arrived, he could see that it wasn't a mountain lion. "God almighty!" he said out loud to himself. "That's a jaguar!"

Determined to document the event, Glenn ran back to his mule, radioed his daughter Kelly that the hounds had bayed a jaguar, and grabbed, not his rifle, but a simple point-and-shoot camera. Then he started snapping photos, the first ever of a wild jaguar in the United States. Later he wrote that, "It was the most beautiful creature [he] had ever seen." The big cat had "fire in his eyes." One of the most striking of Glenn's photos shows the jaguar, looking back at Glenn and standing at the edge of a cliff, with mountains, stark and bare-boned, and dusty valleys spreading out behind him. Later Glenn would learn that a Mexican *federale* shot the jaguar just thirty miles south of the U.S. border. By that time, however, the big cat had already changed the course of Glenn's life and had set off a series of small eruptions on both sides of the border about the viability of the jaguar in the northernmost reaches of its range.

Jack Childs's sighting happened just six months later in the rugged Baboquivari Mountains. Childs, an Arizona outdoorsman and retired land surveyor, had been tracking mountain lions on mules across the Sky Islands of southern Arizona for decades. Like Glenn, the Stetson-wearing Childs looked the part: lean, blue-eyed, and mustached.

Early in the morning on August 31, 1996, Childs and his wife Anna Mary left home, met two friends at a rendezvous point, saddled their mounts, and turned their hounds loose. Their idea was to take a leisurely ride and allow the dogs some exercise and a chance to toughen up their feet.

The group set out in the crisp air of the still-dark day. When the sun rose, they could see that the summer monsoon, spilling rain across the desert, had done its work. The creeks ran with water, and the hillsides were covered in lush green brush and wildflowers. About two hours into their ride, the dogs struck a track. The riders followed them through a canyon, but when the track veered up a sharp, rocky ridge, Childs and Anna Mary took a break in the shade of a juniper tree, while their two friends galloped ahead, trailing the frantic hounds. When the baying of the dogs turned to a "rapid, staccato chop," the Childses knew that the hounds had treed a big cat. Not long afterward one of the riders returned, announcing that they had "treed a jaguar."

The manic dogs circled the tree, but the jaguar appeared unconcerned, lying casually on a limb. Childs noticed the jaguar's distended belly, which meant that it had been feeding on a kill.

After half an hour of filming, the first of its kind ever captured on video in the United States, Childs decided it was time to leave the jaguar alone. He later wrote in *Ambushed on the Jaguar Trail* that "God had truly blessed [them] with this chance-of-a-lifetime encounter."[1] Anna Mary wrote that they had all been grateful for the opportunity to "admire one of God's most magnificent and secretive creatures." The details of that morning, she said, would be forever "etched" in their memories.*

PREDATORS AND PIONEERS IN THE American West have had a tense and wary relationship from the moment they met. Not until the birth of the modern environmental movement did the country's attitude and its fraught relationship toward predators begin to change. Bounties in the Southwest were removed in the late 1960s, and in 1969 it became illegal to kill jaguars under an order by the Arizona State Commission. Attitudes regarding endangered species in general were also changing. In 1966 Congress passed the Endangered Species Preservation Act, which gave the secretary of the interior the power to list and protect endangered native species.

Three years later it passed the Endangered Species Conservation Act (ESCA), which added to the original act protective status for species across the world in danger of extinction. For all its merits, ESCA left behind a complicated legacy, at least as far as the jaguar was concerned. It created two lists, one for species native to the United States and one for foreign species. The jaguar was confined to the latter and was listed as endangered only from the U.S.-Mexico border south through Central and South America.

In 1973, however, when President Richard Nixon supported the

---

* Jack Childs's story about his encounter first appeared in the magazine *Full Cry* and was later published by Rio Nuevo in book form.

transformational Endangered Species Act (ESA), the ESA replaced ESCA with one comprehensive list. The jaguar, however, was either clumsily omitted or knowingly left off because of the prevailing belief that jaguars no longer existed in the United States. Five and a half years later the USFWS published a notice accepting responsibility for what it called a "mistake" and indicated that it intended to rectify the problem as soon as possible.[2]

In the ensuing years, the jaguar was again forgotten. Then in 1992 the USFWS received a petition initiated by the wildlife biologist Tony Povilitis and his students at the University of California at Santa Cruz to list the jaguar. According to the USFWS, the petition included convincing information showing that listing needed to be seriously considered. In July 1994 it published a proposed rule to extend endangered species status to the jaguar. But in April 1995 Congress temporarily extinguished any hope of protecting the jaguar by enacting a moratorium on listing actions. That moratorium was lifted just one year later by means of a presidential waiver, and in September 1996 the Center for Biological Diversity filed a lawsuit and motion for summary judgment for the USFWS secretary to finalize the listing for the jaguar.

In January of the following year, the Arizona Game and Fish Department and the New Mexico Department of Game and Fish, hoping to keep jaguar management out of federal hands, requested that the USFWS reopen the jaguar public comment period. Public opposition to the listing generally fell into two dozen categories, the most common of which was the assertion that the jaguar, a "phantom species," was not native to the United States.[3]

The USFWS rebutted that argument, writing, "The fact that individuals occur in the United States warrants their consideration for listing . . . and development of appropriate conservation considerations."[4]

Meanwhile those who supported listing reminded USFWS officials that the ESA required the designation of "critical habitat" for a listed species when "prudent and determinable." In response, the USFWS claimed, paradoxically, that the "designation [of critical habitat] . . . would not be beneficial to the species."[5]

In July 1997, just ten months after Jack and Anna Mary Childs saw and photographed a jaguar, the long and acrimonious dispute over whether to list the jaguar in the United States was finally settled when the USFWS decided, "After a thorough review and consideration of all information available . . . the jaguar should be classified as an endangered species."[6]

Alan Rabinowitz, who had flown down to Arizona with Salisa to ride mules in the backcountry with the Childses and Warner Glenn and assess the habitat, praised the USFWS for listing the jaguar and defended it for not buckling to pressure to establish critical habitat. He explained that for an animal with a discernible presence in the United States, it made sense. But in the case of the jaguar, he argued that even though big cats had recently been spotted in Arizona, there was no native population of jaguars and therefore no landscape critical to its survival.

Just a week after the jaguar was listed under the Endangered Species Act, the Arizona–New Mexico Jaguar Conservation Team (JagCT), an eclectic new organization of governmental agencies, biologists, conservation groups, landowners, and interested citizens, held its second meeting in Lordsburg, New Mexico. Steve Spangle of the USFWS and Terry Johnson, nongame chief and endangered species coordinator for the AZGFD and the JagCT's capable chairman, articulated their agencies' opposition to critical habitat. Johnson explained that the discussion evoked "an immediate negative reaction on the part of any freedom-loving American who does not want to be constrained by government and dictated to by government." He added that, "Regulatory approaches tend to feed existing hostilities."*[7]

Although the JagCT would eventually be undone by gaping ideological divides over critical habitat, early on, the excitement was palpable. Johnson saw himself as a "facilitator" and made it known that he was not there to impose a point of view.[8] People from different backgrounds, with dramatically different perspectives, worked

---

* Johnson added, "If you can conserve a species adequately, without it becoming listed federally, then you can save yourself that regulatory burden of bureaucracy and all of the stuff that comes with it." Quoted in Steller, "Jaguar Team Ceases Work."

together, and a sense of common purpose prevailed. Jack Childs, who had gone to the Brazilian Pantanal to study with Peter Crawshaw, formed the Borderlands Jaguar Detection Project, which operated under the aegis of the JagCT. Childs placed the first motion-sensing cameras in potential jaguar territory in this country.*

In December 2001, Childs and his wife Anna Mary were going through photos, when they came upon an image of a big cat, the first camera trap photo of a wild U.S. jaguar. The Childses named him Macho A, or Male A, and in the ensuing months they got dozens more pictures of him.

It was a thrilling time for the JagCT, especially with the discovery of Macho A. But an inevitable rift arose among those who wanted to aggressively pursue restoration as per the Endangered Species Act and those who wanted to avoid the regulatory approaches that Johnson had alluded to years earlier.

Alan Rabinowitz weighed in on the divide, advocating for a "good sense" conservation approach that wouldn't interfere with the livelihood of human inhabitants, namely ranchers.[9]

The Center for Biological Diversity grew impatient with the JagCT's and the USFWS's tolerance for the status quo and sued the service, demanding that it establish critical habitat and a jaguar recovery plan as the ESA stipulated. A jaguar crossing into the desert Southwest from Mexico, it said, would have to navigate a dangerous patchwork of private, state, county, federal, and Native American lands. But ranchers disagreed; some who were already up in arms over the reintroduction of the Mexican wolf disagreed vehemently. They resented the idea of federal oversight over another predator with a predilection for killing livestock.

At a JagCT meeting in Animas, New Mexico's Warner Glenn, who had developed a reputation for centrism, spoke, saying that cattlemen felt betrayed by the lawsuit. They worried that their right

---

* Childs returned to Arizona from Brazil with six hours of video, some of which he shared with the larger JagCT group. That same year he published *Tracking the Felids of the Borderlands*, a field guide to the various wild cats native to the Southwest.

to use their own property could be curtailed by efforts to expand and protect jaguar habitat. Some JagCT members, and Rabinowitz, too, feared that any effort to assign critical habitat might backfire, causing mistrustful cattlemen to kill a jaguar rather than report its presence.[10]

Though in its ground rules, the JagCT stressed the importance of holding "professional and cordial" meetings, the rift among its members grew increasingly quarrelsome. This was especially so after February 2006, when Warner Glenn and his hounds discovered another jaguar in southwestern New Mexico's Animas Mountains. Once again Glenn left his rifle in its scabbard and photographed the big cat. Appropriately, Glenn named the handsome jaguar Border King.

Just months after Glenn's sighting, JagCT members discussed capturing and collaring a jaguar. In its minutes, the JagCT noted that the USFWS agreed with the plan but would not declare its support publicly. The minutes also said that the USFWS's only caveat was that it demanded a thorough capture protocol.

In principle, everyone agreed that collars offered an abundance of information, namely the identification of crucial habitat and travel corridors. But some argued that the risk of catching and sedating, say, Border King, who may have been the country's only wild-roaming jaguar, carried far too many risks. Opponents also said it didn't make sense to risk putting a GPS collar on a jaguar because state and federal officials had already declared their unwillingness to use the data to protect key habitat.*

Howard Quigley and Brian Miller, two members of the scientific advisory committee, wrote a scathing letter accusing the JagCT of being bogged down in fruitless personal and political debates. They advocated for the "'best available science'" and stressed that

---

* Michael Robinson invited Samuel Wasser, who directed the University of Washington's Center for Conservation Biology, to talk with the JagCT about scat-sniffing dogs and the information that could be derived from scat. Melanie Culver at the University of Arizona was capable of conducting mitochondrial scat analysis in her lab. "Once the discussion started toward collaring, [the JagCT] immediately split into two factions," said Shiloh Walkosak, a volunteer for the Borderlands Jaguar Detection Project.

the JagCT needed to avoid "parochial" squabbles over collaring and critical habitat.¹¹

Meanwhile Jack Childs continued his work on the Borderlands Jaguar Detection Project. While examining the spot patterns on various photographed jaguars, he came upon an exciting discovery: The jaguar that his cameras had captured in December 2004 was the same one his dogs had treed in the Baboquivari Mountains in 1996. He identified the big cat by its distinctive rosettes, one on his lower right shoulder, near his ribcage, which resembled the character Pinocchio, and one on his lower left shoulder that looked like the animated cartoon character Betty Boop. Childs estimated the jaguar's age to be ten years old and dubbed him Macho B.

Childs and his young grad student assistant, Emil McCain (who had studied in Costa Rica with Eduardo Carrillo), continued to install cameras in the rugged terrain north of the U.S.-Mexico border. McCain, an excellent tracker who believed that jaguars had never left Arizona and had only become more furtive, spent long days on foot, recovering images from thirty-two cameras.* Based on those photos, Childs and McCain determined that Macho B's range exceeded five hundred square miles. He was a prodigious wanderer, triggering cameras a dozen miles apart within a span of just hours. Childs and McCain used these photos as proof that Macho B was a mature, resident jaguar, "well beyond the age of dispersal."¹²

In 2008 Jack Childs and Emil McCain authored a contentious article, writing that they had "videotaped several scent-marking behaviors, indicating the residency of adult jaguars within Arizona." Using historical records, the two authors also pointed out that "12% of all known jaguars in Arizona were females raising young, clearly representing a breeding population north of the border."†¹³

---

* McCain believed jaguars had "always been" there and that prior to 1950, sightings were frequent because "cowboys . . . prospectors . . . [and] homesteaders . . . traveled more often into remote areas." Mahler, *Jaguar's Shadow*, 205–6.

† Later, in an interview with me, Jack Childs distinguished between residents and "nonpermanent residents." He also said that a "breeding population" was not necessarily a "sustainable population."

Childs was now caught in the middle of an ugly argument. Despite his paper, he opposed a federally imposed critical habitat designation. He also abhorred what he considered the center's belligerent tactics. A consensus builder, he trusted in the JagCT's model for jaguar conservation, which relied on the involvement of state agencies, voluntary local collaboration, and an advisory document that he believed could unite divergent interests in the Southwest. However, internal disputes at the JagCT continued. Members considered leaving over what they saw as the organization's obeisance to livestock interests and state officials who were reluctant to put jaguar recovery in the Southwest in federal hands. They protested the team's continued opposition to protecting jaguar habitat. Some took to calling the fractious group the Jaguar "Conversation" Team and accused members of spending little time actually doing the hard work of restoring an endangered species.

According to Tony Povilitis, "There was never clear talk about jaguar recovery. It was called jaguar conservation, not recovery. . . . Conservation can be narrowly interpreted, and in this case it was."[14]

But what would ultimately undo the group happened some two years later on February 18, 2009, when the aging Macho B, lured by the scat sample of a captive female jaguar in heat, stepped into a snare attached to a mesquite tree in the Atascosa Mountains of Arizona.

CHAPTER 27

The Tragic Cat

ON THE CHILLY MORNING OF FEBRUARY 18, 2009, TWO AZGFD biologists were tracking Macho B, following his pug prints across the canyon floor. Just weeks prior, Emil McCain had led the same biologists on a tour of the snare lines. Under McCain's instructions, the snares had been set and baited with jaguar scat. McCain, however, warned the biologists that it was a "hush-hush" operation that had been approved by people in charge at the USFWS and the AZGFD.[1]

Near midmorning, at four thousand feet, the biologists found Macho B in a snare. Once caught, he had laid waste to everything within reach. When the biologists approached, he lay passively on the frosty ground. But when they darted him, he again turned aggressive, jumping to his feet and swatting the air before the Telazol could take effect.

The biologists waited until they were certain Macho B was asleep, then put liquid tears in his eyes and placed a bag over his head. First, they took his vital signs. When his anal temperature reading came in at 94.8 degrees, they grew anxious. The big cat had been lying on the cold ground and was nearly hypothermic, and neither biologist possessed the medical training needed to address Macho B's postcapture trauma. To make matters worse, the biologists had no idea how long

Macho B had been in the trap because the snare lacked a monitor indicating when it was triggered.

The pair dragged the jaguar into a sunny area, laid his body on top of a sweatshirt, and tested his vital signs. They collected DNA samples (hair, swab, and blood) and scat samples from his colon. Next, they inspected his teeth and noted that some were missing or broken. They weighed him with a portable scale; he weighed 118 pounds (120 pounds was considered average for an adult male "northern" borderland jaguar), and they noted that his belly was full. They sprayed iodine on the leg wounds he'd sustained while in the snare. The last thing they did was to fit him with a one-pound, twelve-ounce satellite collar, programmed to deliver data downloads every three hours. Then they watched and waited. At three p.m., six hours after they first darted him, Macho B, still groggy from the effects of the Telazol, struggled unsteadily to his feet, gave a "deep, throaty growl," and staggered off.[2]

Macho B's capture fulfilled a goal that the AZGFD and some JagCT members had been harboring almost since its inception, but especially since 2008, when the Department of Homeland Security released a $50 million plan to analyze and limit the border fence's effect on endangered species like the jaguar. Following the announcement, the AZGFD and the JagCT traded enthusiastic messages, suggesting that the collaring of a jaguar could create promising funding opportunities.

When the two biologists reported Macho B's capture to AZGFD headquarters, people involved in the program celebrated—but quietly, until a day later, when Terry Johnson announced the historic event and AZGFD officials issued a press release: The capture and collaring of Macho B would offer AZGFD authorities the best information on jaguar behavior in the United States. Those insights would aid the federal government as it assessed its border development projects and their effects on endangered species. The press release perpetuated the narrative that Macho B had been caught fortuitously—in an area where AZGFD researchers were conducting a comprehensive study of wildlife movement corridors. The story worked for a few days, but not long after Macho B's GPS collar failed to transmit a signal, things inside AZGFD began to unravel.

The *Arizona Daily Star* reporter Tony Davis and fellow reporter Tim Steller had doggedly followed the story of Macho B's capture, including filing a Freedom of Information Act request. On February 21, three days after Macho B was captured, Davis interviewed AZGFD officials who stuck with the bewildering organizational deception that Macho B's capture had been incidental and had taken place in an area where Macho B had not been spotted before. "We were trapping lions and bears," one official said, "and we prepared for the possibility of capturing a jaguar."[3] Another reiterated the party line: "We knew there was potential to ensnare a jaguar. That was not the purpose or intent."[4]

Inside the USFWS, a fervid controversy was brewing. Internal discussions revolved around AZGFD's agreement with the USFWS and whether the word *jaguar* was ever explicitly mentioned in its 1998 and 2007 "incidental take" permits. Then there was the question of whether the jaguar handling protocol had ever been accepted by the USFWS.

Meanwhile McCain, who was tracking Macho B via GPS, reported to the AZDGF that the big cat had been stationary for twenty-four hours. Worried that the collar had malfunctioned, McCain emailed the CEO of the company that had manufactured it, asking how to interpret the most recent data.

On March 1, nearly two weeks after Macho B was collared, the AZGFD sent out a search party. The team successfully located the big cat. According to an AZGFD biologist, he appeared "weak and wobbly . . . like he was in a stupor."[5]

A day later the AZGFD's story about the capture disintegrated. Janay Brun, McCain's and Childs's assistant, guilt-ridden by her role in capturing Macho B., including filing a Freedom of Information Act request,* emerged as a lone whistleblower. Brun disclosed that she had placed the jaguar scat at the snare site at McCain's instruction. McCain, she said, had repeatedly reassured her, "Everything's cool, everything's legal."[6]

---

* Davis and Steller would eventually file a Freedom of Information Act request.

On the same day, a team—including AZGFD employees, a consulting veterinarian, a houndsman from Wildlife Services and his dogs, and a helicopter pilot—set out to find and capture Macho B. Once the dogs struck Macho B's track, it didn't take long for them to bay the jaguar and for the vet to place his dart.

Late that afternoon they transported Macho B to a treatment room at the Phoenix Zoo. The big cat was feverish and frail—it had lost almost twenty pounds in two weeks. Thirty minutes later the lead veterinarian at the zoo was able assess the results of Macho B's blood test. His kidney readings were dangerously high.

The zoo's veterinarian considered treating Macho B with fluids, but decided against it because it would require administering sedatives that could batter the cat's kidneys. The vet considered a kidney transplant, then vetoed that idea. Another option was to let Macho B die a natural death. That idea was similarly dismissed. Ultimately it was Steve Spangle, USFWS field supervisor for Arizona Ecological Services, who made the wretched decision to euthanize Macho B.

On March 3 local and national newspapers covered Macho B's death. Under a headline that read NO JAGUARS LEFT IN U.S. AFTER DEATH OF MACHO B, the *Arizona Capitol Times* wrote, "Kidney failure brought on by age—likely aggravated by the stress of capture—led to the death of the only known wild jaguar in the United States."[7] The *Los Angeles Times* hinted at the irony of the situation in an article headlined ARIZONA JAGUAR'S DEATH PROBABLY HASTENED BY CAPTURE, ZOO VETERINARIAN SAYS.

Two days after Macho B's death, the Center for Biological Diversity called for an investigation. While members of the USFWS, the AZGFD, and the Phoenix Zoo gathered for a press conference to brief the public and media outlets on the probable causes of Macho B's death, the Center held a memorial service for the big cat outside the USFWS's Tucson office.

Within two weeks of his death, Macho B was skinned, and his pelt was sent to a tanner. In the meantime, his skeleton and skull were cleaned by a flesh-eating beetle colony at the San Diego Natural History Museum. For genetic purposes, samples of his muscle tissue were filed away at the Endangered Species Center at the San Diego Zoo. His blood, hair, and fecal matter samples were stored at the Cooperative Fish and Wildlife Unit at the University of Arizona. Still more samples were sent to the USGS Wildlife Health Center in Madison, Wisconsin, and to the University of California at Davis.

In its March 16 report to the JagCT, Terry Johnson and William Van Pelt of the AZGFD and James Stuart of the New Mexico Department of Game and Fish expressed their exasperation over the negative turn that press coverage of the Macho B incident had taken. Much of it, they wrote, was flawed and overwhelmingly biased.

On March 29 Tony Davis published an explosive piece in the *Arizona Daily Star*, titled "Did Jaguar Macho B Have to Die?" The article was accompanied by a photo of Macho B lying on the ground with his left front foot caught in a snare. Davis wrote, "Sorting truth from opinion will be difficult because officials chose to perform a 'cosmetic' necropsy rather than a full one."[8]

Indeed, the zoo had conducted the less invasive procedure at the request of officials at the USFWS and the AZGFD who wanted to leave the skin intact to create a "live mount" to be exhibited for educational purposes. A complete necropsy would likely have provided better information, but in that event Macho B's skin could not have been salvaged. The zoo also chose not to take tissue samples of the brain and spinal cord, which might have helped to explain Macho B's condition at the time of his death.

On April 1, 2009, the USFWS Office of Law Enforcement and the U.S. Department of Justice launched a criminal investigation into Macho B's death. Its two investigators scoured USFWS, AZGFD, and JagCT documents and emails and ultimately conducted thirty-eight interviews.

Late that year, U.S. district judge John Roll ruled that the USFWS had failed to employ the best scientific evidence available when it decided that the jaguar did not require critical habitat in the United States. Furthermore, Judge Roll ordered Fish and Wildlife to review his ruling and make a decision by January 8, 2010, on designating critical habitat and preparing a jaguar recovery plan. The

Center for Biological Diversity's Michael Robinson, who had worked on the JagCT's habitat subcommittee, celebrated the judge's decision. "Denying the jaguar protection because it is overly endangered is an oxymoron," he said. "That was the essence of the government's plan, that there are so few jaguars that they don't need a recovery plan. And the judge saw right through that."[9]

Just a month after Judge Roll's order, Tim Steller reported that the JagCT had ceased its activities. According to Warren "Bud" Starnes, a policy specialist for the New Mexico Department of Agriculture, it was the environmentalists who sank the JagCT. "It had very laudable objectives," he said. "But the enviros started pressing, trying to get maps of habitat. Then they started threatening lawsuits." Terry Johnson, regretting the divisions that had riven the JagCT, summed it up this way: "It's really tough to operate somewhat in the center—not necessarily straight down the middle, but to borrow the best from the left and the best from the right . . . and try to develop that magic concoction that ultimately works to benefit the jaguar and the people."[10]

Tony Povilitis had a distinctly different take. "As the years went on," he explained, "there was more and more resistance to doing the habitat conservation work, to the point where essentially nothing got done." Michael Robinson was equally critical. "At every juncture, there was an effort to limit the scope, the scale and impact of any kind of action to preserve jaguar habitat," he said. Sergio Avila, of the Sky Island Alliance, also weighed in on the fatal rift, describing how the team had never fully accepted the jaguar as an endangered species in the United States and, consequently, never embraced the idea that arriving at an acceptable jaguar recovery plan was its primary task.[11]

Just months later the Obama administration announced that it would protect the endangered jaguar's prime U.S. habitat and simultaneously develop a jaguar recovery plan. The New Mexico and Arizona Cattle Growers' Association promptly criticized the decision.

Meanwhile in the desert Southwest, the Macho B story would not die. Many biologists across the corridor viewed the hoopla regarding Macho B, a lone geriatric cat, as a distraction, especially when jaguars were being shot and trapped and poisoned every day in Mexico and in

Central and South America. They believed it complicated their conservation efforts by evoking unnecessary worries about the safety of capture and collar projects, which they regularly conducted under the most stringent protocols. One biologist confessed that "the mystique of the Macho B story didn't do shit for [him]." Another replied that it was a "big effing mess from the start" that grew progressively worse because of the obfuscation of state and federal agencies.

After months of yeomanlike work, in which they waded through fourteen thousand pages of evidence and testimony, Tony Davis and fellow reporter Tim Steller confirmed some of the case's uglier accusations. In a January 21, 2010, article subtitled "State's Capture of Jaguar Macho B Was Intentional, Federal Investigators Conclude," they laid out those facts for everyone to read. The capture of Macho B was intentional, according to a new investigative report by the Interior Department's office of inspector general. Moreover the AZGFD had been aware that Macho B was near a site where department employees had set traps, and it had failed to consult with the USFWS about the jaguar's presence, as required by federal law. Finally, the report asserted that the USFWS wrongly approved a cosmetic necropsy.[12] Ultimately, the investigation would find that numerous violations of the Endangered Species Act were committed.*

Just over a week before the explosive article by Davis and Steller, the USFWS announced that "new data" made it "prudent" to designate critical habitat in the United States for the jaguar. Two weeks later, in an angry and incredulous op-ed, Alan Rabinowitz declared it a "shocking" turn of events and wrote that there had been nothing indicating that the desert Southwest had become part of a normal

---

* Michael Robinson laid out what the ESA would have required of AZGFD: "In the event that an endangered species stands the possibility of being incidentally captured, Section 7 of the ESA directs nonfederal entities to contact the appropriate local USFWS office for a biological opinion and initiate the Section 10 'incidental take permit process.' The USFWS then conducts its own biological assessment, and the applying entity can either be issued a Section 10(a)(1)(A) permit, which authorizes the direct take of an endangered species, or a Section 10(a)(1)(B) permit, which authorizes the incidental take of an endangered species provided that the action involved is incidental to an otherwise lawful activity." Robinson, "Investigative Report on Macho B."

ranging pattern for jaguars. He insinuated that the USFWS had not based its decision on hard science but rather had bent to the winds of change and had been bullied by the unremitting legal pressure of the Center for Biological Diversity, Defenders of Wildlife, and the Macho B debacle.[13]

Adding that jaguars had failed over the last century to reestablish themselves in the American Southwest, he suggested that the job would have to be undertaken by the government, at considerable expense. He reiterated a point he'd made on many other occasions: conservation efforts and limited monies should be focused "south of the border" on the Jaguar Corridor, where "thousands of jaguars live and breed in their true critical habitat." He argued, too, that if "critical habitat is redefined as any place where a species might ever have existed, and where you or I might want it to exist again, then the door is open for many other senseless efforts to bring back long-lost creatures." Later, in *An Indomitable Beast*, he would write of the irony: "I found myself in a strange and unlikely position: fighting against critical-habitat designation for a species that I had spent most of my professional career trying to save."[14]

Just two months after Rabinowitz's scathing article, the Center for Biological Diversity asked the federal government to set aside over 53 million acres for the jaguar across New Mexico, Arizona, southern California, and West Texas, an area of the Southwest more than half the size of California, claiming the species needed huge areas in which to roam. Many, including those otherwise sympathetic to the jaguar's critical habitat needs, were shocked by the enormity of the request (six times the size of the critical habitat set aside for the Mexican spotted owl).

In May 2011 the two-year-long criminal investigation of the capture and death of Macho B ended. Emil McCain was convicted of a misdemeanor for the taking of an endangered animal, was fined $1,000, and was banned from doing jaguar work for five years. Janay Brun, who in her book rebuked members of the AZDGF and the Borderlands Jaguar Detection Project, took a plea deal whereby criminal charges were dismissed in return for her admission that there had been no "authorization or permission" to capture jaguar Macho B and for her promise that she would not participate in any big cat research for one year. The U.S. Attorney's Office official spokesman revealed that no criminal charges were planned against anyone else. Five AZGFD employees were referred for prosecution, but no charges were filed. Jack Childs would lament that the Macho B incident destroyed the credibility of the Borderlands Jaguar Detection Project and sent jaguar research in the Southwest "back to the stone age." According to Chris Bugbee, who had tracked El Jefe for years, the Macho B debacle revealed the high-stakes nature of big carnivore studies. "They bring out the carnivore in scientists," he said.

For Gary Hovatter, the AZGFD's deputy director, the whole imbroglio was full of lessons about "personalities and motivations, ego and ambition."[15]

Meanwhile Alan Rabinowitz announced that he and his team, after five years of exhausting work, had managed to finish ground-truthing and mapping the Mexican and Central American Jaguar Corridor. The borderland region of Arizona and New Mexico was omitted from those maps.

Rabinowitz and Howard Quigley flew to Bogotá, where they met with Colombian vice president Francisco Santos, the minister of environment, and the director-general of national parks to discuss a critical pathway to the Colombian Amazon. The three men signed the agreement recognizing the corridor and pledging themselves to preserving it.

As Rabinowitz and Quigley continued securing commitments from various governments along the corridor, Macho B's tragic death was still reverberating across the American Southwest. In December 2012 *The Arizona Republic* ran an award-winning three-part series by Dennis Wagner titled "Macho B: Last Roar of the Jaguar," in which the journalist suggested that because environmentalists had been opposed to the notion of capturing and collaring a jaguar, the effort to capture Macho B had to be undertaken surreptitiously.

Like Davis and Steller, Wagner filed numerous public record requests and obtained thousands of pages of documents for his report.

He also gave Emil McCain an opportunity to defend himself and wrote that McCain had been portrayed inaccurately as a "rogue scientist who duped negligent government bureaucrats." The truth, he wrote, was that "Macho B fell victim to environmental politics, greed and a quest for federal Department of Homeland Security money."[16]

Despite the death of Macho B, Jack Childs couldn't just walk away from the jaguar research he had once loved. He helped write the proposal that the University of Arizona Wild Cat Research and Conservation Center submitted to the Department of Homeland Security. He remained active in a "citizen science" program run by Melanie Culver of the University of Arizona and in a research program south of the border.

But it was Janay Brun who captured the whole unseemly Macho B mess. In her book *Cloak and Jaguar*, she wrote poignantly of the big cat's death, adding, "It can't be stressed enough that . . . Macho B was the nation's only known wild jaguar." [17]

And then he was gone.*

---

* "In March of that year Arizona's 'resident' jaguar, named Macho B, was killed by those who were tasked with protecting him. His death was inadvertent, but avoidable. Lured into a humane trap by biologists who did not have the authority to take this action, he was examined, collared, and released. He died a week later, largely from the effects of the stress associated with his capture." Wilcox, "Encountering El Tigre," 234.

CHAPTER 28

Land of the Jaguar

AN HOUR BEFORE SUNSET, THE HEAT WAS STILL PUNISHing as I had ascended the steep, cut-stone steps of Temple I, Calakmul's second-highest structure. Calakmul's directors had given me special permission to remain at the ruins after the park closed to the public. Sunset seemed a fitting and symbolic time to contemplate an extraordinary Maya society that worshipped the jaguar and flourished for almost fifteen hundred years. When archaeologists excavated Calakmul in the 1930s, they uncovered the tomb of its greatest ruler, known as Jaguar Claw.

I reached the top of Temple I and stood in the middle of a rectangle of stones. While I waited for the sun to begin its precipitous dip, I wondered what it would be like to hide in the temple until dark and watch the stars. Envisioning the sky as a stage, the Maya made apparent the story of creation so that all its people could embrace its power and truth. At the base of the temple, on the platform that supported the pyramid, lay three carefully arranged altars associated with the constellation Orion, which according to the Maya was the birthplace of the gods. Researchers suggest the configuration represents a celestial hearth, at the center of the universe, containing the first fire of creation. The stones, the Maya believed, were carried in a canoe, propelled by jaguar and stingray paddlers, who followed the Milky Way to the locus of Creation. Even today, when the Maya build cooking fires, they use three stones, arranged to form a triangle, to imitate what Jim Reed of the Institute of Maya Studies calls the three "hearth stars" of Orion that surround the supernatural fire.

As I gazed out at the ancient Maya site below, one that covers seven square miles and contains over six thousand structures, I imagined a shaman or a priest, carrying a ceramic jar of incense made from the sap of the copal tree, climbing the steps of Temple II, the "Great Pyramid," to retrieve a message from the gods. Among the Maya, priests and shamans were capable of achieving a state of grace that enabled them to move between the ordinary world and the realm of the gods, acting as human agents of these holy messages.

Calakmul, now a World Heritage site, was once one of the largest and most dominant city-states in the Maya world. With an estimated population of over fifty thousand people, it brutally imposed political control over a million inhabitants. Its major competitor, Tikal, lay sixty miles due south as the crow flies, across millions of acres of prime jaguar habitat, in the Petén region of northern Guatemala.

The Calakmul ruins are part of the huge 1.8-million-acre Calakmul Biosphere Reserve and its UNESCO-protected forest. A twenty-year-old study of the reserve by Dr. Gerardo Ceballos—a senior researcher at the Institute of Ecology at the Universidad Nacional Autónoma de México, president of the National Alliance for Jaguar Conservation, and perhaps the country's premier jaguar conservationist—estimated that Calakmul was home to 181 to 482 jaguars. In 2016 he updated that survey. He and his fellow researchers discovered that Calakmul's big cat population had increased to six hundred jaguars and that the reserve had a population density of six jaguars per square kilometer.[*] The biosphere reserve comprises a portion of one of the most important jungle habitats in the northern hemisphere, the Selva Maya, a

---

[*] In 2010 Ceballos participated in a national jaguar census—"the only one ever carried out by one nation at the national level"—that estimated Mexico's jaguar population at four thousand big cats.

forest region extending from southeastern Mexico to northern Guatemala and Belize. At 10 million acres, the Selva Maya contains one of the largest contiguous blocks of jaguar habitat north of the Amazon. It has been called an "ark of biodiversity" with some three hundred species of trees and eighty species of mammals. The Selva Maya, though, is in grave danger. Between 2000 and 2015, the forest, once a sea of green, decreased by 25 percent because of fires, cattle ranching, the encroachment of people, and narcotrafficking. The WCS hopes to reverse that trend with its Five Great Forests of Mesoamerica initiative, which aims to protect the entirety of the Selva Maya.*

As the daylight faded in a fleeting purity of color, enhanced by the absence of any artificial light, I spotted a big ceiba tree lifting its great head above the surrounding forest. The ceiba was sacred to the Maya. They believed it stood at the Earth's geographic center. While the bloodred sun clung to the horizon—the Maya believed that at night the sun hid in the belly of a jaguar—and illuminated the top of the Great Pyramid, I spotted a black vulture (*ch'om* in Maya) soaring overhead in search of carrion. Soon cicadas droned, bats screeched, and red-eyed tree frogs began to sing. Not long afterward fierce-sounding roars began in the valley below, as a group of black howler monkeys engaged in their boisterous end-of-day aspirations. For a brief moment, I imagined that the sound might be the coughing, or "sawing," of a jaguar.

When darkness descended and I left the park, one of the guards, a Maya man, told me a striking story. He said, with unmistakable reverence, that during Covid, in a kind of convergence of the spiritual and the material, he watched jaguars wander among the empty Calakmul ruins, as if they had returned to their ancestral home.

THE NEXT MORNING, when our small group gathered in the lobby of our Casa Ka'an eco-lodge, my pants were still dusted white with

---

* In addition to saving the Selva Maya, the WCS hopes to preserve La Moskitia in Nicaragua and Honduras; Indio Maíz-Tortuguero in Nicaragua and Costa Rica; La Amistad in Costa Rica and Panama; and El Darién in Panama and Colombia.

the powder of Calakmul's limestone. Howard Quigley, who was known as "Papa Jaguar" throughout Mesoamerica, announced the day's schedule.

I had been talking with Quigley for the past two years and had been eager to meet him. Over the phone and on Zoom calls, he had been kind, encouraging, generous with his time, enthusiastic about my book, and inordinately modest about his contributions to wildlife science. Every time I tried to steer our conversation toward his career, he would say that we'd have ample time to get to that. He had intended to join us for the capture and collaring campaign in the Pantanal, where he had spent his formative years as a jaguar researcher, but had to cancel because of ongoing health issues.

In the lobby, I also met Diana Friedeberg, the director of Panthera's Mexico operation. Bright and energetic, Friedeberg was one of Panthera's formidable female recruits. As a young woman, she had dreamed of being an artist but abandoned that dream when she entered a Ph.D. program in neurobiology at Duke University. Eventually, after another change of heart, she left the program and returned to Mexico, her home country, to dedicate herself to conservation. Because Mexico represented the northernmost part of the jaguar's range, where populations existed in a semifragile state, she knew that her country was especially important to the success of the Jaguar Corridor. So when she heard that Panthera was hoping to open a Mexico office, she interviewed for the job. She and Quigley met via Zoom, and with his endorsement, she flew to New York to meet Alan Rabinowitz, who offered her the start-up position. Since then she has pursued her Ph.D. in biology and is dedicating her dissertation to Rabinowitz.

Friedeberg had tackled a number of jaguar related issues, but the one that she'd been most concerned with was deforestation and its effects on the jaguar population in the western Yucatán Peninsula. She mentioned one female jaguar in particular—Nicte—that her team had been following for years in the Laguna de Términos Conservation Unit outside the city of Campeche, not far from where we'd gathered.

Nicte was a handsome female, dense with muscle. Her coat was a work of art, highlighted by broken rosettes that turned into dark inky blotches on her legs. She was also a masterful hunter.

In 2016 a camera trap caught her with a cub. Nicte was secretive by nature, but while protecting her cub, she became even more wary of people and roads and camera traps. Later that year, however, coinciding with the arrival of a new Mennonite community, her detection frequency abruptly rose, so Friedeberg grew concerned. According to her field notes, Nicte was appearing more regularly on camera traps, which meant that she had been displaced from her home turf and was searching anxiously for new territory.

Global Forest Watch confirmed Friedeberg's fears. In 2017 alone the state of Campeche lost 173,000 acres of forest cover, with much of the deforestation occurring in the areas where Mennonites had been creating new communities and expanding their landholdings. They razed the forests and replaced them with fields of sugarcane, sorghum, and genetically modified soybeans (transgenic corn is still illegal in Mexico) that they sprayed liberally with pesticides and powerful herbicides. Local Maya, who had been tending beehives for centuries, complained that their bees were dying and their Melipona honey, much of which they sold to the European Union, was contaminated.[1]

The Laguna de Términos Conservation Unit was established to protect the wetlands in the region of the Gulf of Mexico. Together with the Centla Biosphere Reserve, it forms the most important coastal ecological unit in Mesoamerica and protects prime jaguar habitat. But in the last fifty years, it has lost 12 percent of its mangroves and 31 percent of its lowland tropical forest cover. The five-year destruction reveals an even bleaker picture. According to a recent paper authored by Friedeberg, forest fragmentation represents a threat to the survival of the small but still stable jaguar population in southeastern Mexico.

QUIGLEY AND FRIEDEBERG HAD INVITED me to join them and their team on this, the first leg of what Panthera was calling the resumption of Journey of the Jaguar (JotJ). The JotJ was something that Quigley and Alan Rabinowitz had conceived of in 2015. In many ways, they envisioned it as a three-year exploration of the corridor and an adventure in storytelling. But Rabinowitz grew seriously ill, and then Covid with its travel restrictions hit, and Quigley put the journey on an indeterminate hold. Our trip to Mexico's Maya country, almost three years later, was the journey's resurrection.

When Quigley climbed to the top of Calakmul's Great Pyramid, he was profoundly affected by the experience. Flush with excitement, he told me soon afterward, in his basso profundo voice that was made for narrating wildlife documentaries, "Most landscapes are human dominated. That's where we do the bulk of our work. But this one is special; it's one of those rare places." Then he added beatifically, "For there to be jaguar habitat, as far as the eye can see, is inspiring. It warms a jaguar biologist's heart. It gives one hope for the species."

But Mexico, in Quigley's estimation, was a "complicated place for jaguar conservation." Three decades prior the jaguar had been nearly erased from its landscape. Since then it had made a remarkable recovery, particularly in Calakmul. But that recovery was now being threatened by what was arguably the most controversial development in the northern portion of the Jaguar Corridor in decades and one of the largest infrastructure projects in Mexico's history. The $15 billion Maya Train (Tren Maya in Spanish), initiated by the populist president Andrés Manuel López Obrador (AMLO), ranks second only to the U.S. Border Wall and the Panama Canal in its grandiosity and its potential for environmental mayhem. It is forest fragmentation on steroids. However, unlike the Border Wall, which was expanded to keep people out, Tren Maya proposes to bring in hundreds of thousands of eager tourists every year to marvel at the cynosures of the ancient Maya world.

On December 1, 2018, not long after Alan Rabinowitz's death, AMLO was sworn in as Mexico's president. At Mexico City's main square, the Zócalo, which was once the official center in the Aztec city of Tenochtitlán, he participated in an Indigenous ceremony in which he received a traditional cleansing. Native healers brushed him with herbs, and in a ritual of purification, they blew puffs of copal, traditional incense smoke, over him. The Indigenous activist Carmen Santiago Alonso then handed the president a curved ceremonial staff, made of cedar wood and engraved with the president's name and the shape of the Mexican eagle as a symbol of AMLO's commitment to Mexico's Indigenous communities.

It was the first time in the country's history that a Mexican president had ever taken part in a ceremonial inauguration by Indigenous groups. In return for their allegiance, AMLO promised the assembled crowd that Indigenous peoples in Mexico would receive priority in his government's social programs. As a man who hailed from the poor and undeveloped south, he told them he understood their long-lived pain and frustration. He promised that the plan for the train, which he had just unveiled, would "spread the wealth" to overlooked portions of the country by carrying vacationers from resorts in Cancún and Tulum into the jungle. He characterized the $10 billion project as a restorative "act of justice." It was a message that people in Mexico's south, tired of false hope, greeted enthusiastically.[2]

The train will travel over 966 miles through the states of Quintana Roo, Yucatán, Tabasco, Campeche, and Chiapas, and large stretches of pristine jaguar habitat, triggering mass development throughout the region. For Mexico's jaguars, Tren Maya is devastating news. Mexico contains 69 percent of the Central American corridors in terms of total area, with a mere 3 percent of that area protected.

Quigley didn't parse his words when it came to the train and its impact. He called it "Mesoamerica's largest and potentially most destructive project in many, many years," adding that sweeping transportation projects often spell doom for animals. For jaguars, he said, they are potentially "the first step toward extinction."

CHAPTER 29

AMLO's Iron Horse

AFTER OUR SOJOURN IN CALAKMUL, WE TRAVELED TO the scenic village of Ejido Miguel Colorado. Miguel Colorado is the kind of place where one half-expects to see an old Maya woman, walking at the edge of a dusty road, dressed in a colorful *huipil*, carrying a small bundle of firewood or pumpkins, returning from her milpa. It is also home to eight jaguars, and though the land is unprotected, it is connected to the Gran Calakmul region and is likely the primary source of jaguars that pass through the corridor linking Laguna de Términos with the Calakmul Jaguar Conservation Unit.[*]

Alan Rabinowitz had been here, too, not long before he died. When he visited, the people of the community butchered and cooked chickens in his honor, and an old man of the village told him a parable, the lesson of which was that the forest was like a life-giving house, and the animals of the forest held up the house. However, the old

---

[*] A team of biologists who documented eight jaguars in Miguel Colorado urged that the "integrity of the Miguel Colorado area should be preserved through efforts to protect both the habitat and the jaguars living there." Hidalgo-Mihart, Contreras-Moreno, et al., "Validation."

man cautioned, if the animals disappeared, the house would collapse, just as a forest would. Rabinowitz never forgot the story.

Just outside the village, one could see that the symbolic house the old man had invoked was falling down. Bulldozers were clearing a forty-meter-wide swath of railbed through what had once been untouched rainforest. We visited Sergio, a communal peasant, one of Panthera's Indigenous workers participating in a kind of "guardian stewardship" program in which locals take part in wildlife conservation by attending to camera traps. In his rustic home, we heard that the train had the energetic backing of a large portion of the local Maya population.

Sitting in his pretty backyard with wandering dogs and a snorting caged pig, and surrounded by homemade shelves decorated with pots of flowers and pepper plants, Sergio spoke of the train. He was a heavily muscled man who had obviously spent much of his life doing backbreaking manual labor, which had afforded him a tidy two-room house. The train, which would deliver tourists from the beaches of Cancún, Playa del Carmen, and Tulum, offered him, and the people of his community, hope. Sergio's son Gabriel, however, offered a more nuanced opinion of the train. Young men had jobs, he said gratefully, and AMLO had promised better education, health care, and housing, but the animals he used to see at the edge of the forest had fled to escape the construction crews and commotion. In order to bring in throngs of tourists to goggle at the beauty of the Maya world, the Mexican government and the men who filled the work crews had to destroy ancient Maya forests and the animal kingdom they supported.

Paul Worley, a professor of English at Western North Carolina University, who has published two books on Maya literature and has spent considerable time among the Maya, captured the duality of the train, calling it a "blind alley." He explained that the Maya have few options. "They would rather their loved ones work in the tourist industry in Cancún or for the train than head north to the United States. The train will allow them to stay on their land and make some money at the same time, because in many ways, the milpa that has sustained Maya communities is no longer economically feasible, and with prolonged drought, it's no longer environmentally feasible."

The following day at Chetumal ruins, near the border of Belize, on the southeastern tip of the Yucatán Peninsula, we met another determined supporter of the train, Eleazar Dzib, a spokesman for a group of sixty Maya villages. Dzib said that the Maya of the Calakmul area were proud and spirited descendants, Cruzob, of defiant Maya who had staged a prolonged rebellion against their Spanish oppressors.

Stout with broad shoulders and high cheekbones, Dzib was full-blooded Maya on both sides of his family. He spoke with a quiet dignity, explaining that he was a newly elected representative of a group of Maya *ejidos* in the Calakmul area. Like many modern Maya, he inhabited a fragile existence between the ancient and the modern, embracing contemporary conveniences while residing culturally in a world his ancestors would recognize.

He told us it was his job to guard the interests of his people. Like so many Maya, he believed his country had willfully neglected his people. But AMLO was committed to changing that. Tren Maya made the promise real.

Eventually, as Friedeberg turned the discussion to jaguars, Dzib told us that for countless centuries, his people had had to live as neighbors with the powerful and enigmatic cats, and they did so by acknowledging that part of the world belonged to the jaguar. The modern Maya, though long separated from the jaguar myths that had once animated their people's spiritual and everyday life, still respected the jaguar's ferocity. In fact, with regard to the train, he said, the Maya would protect what was theirs with the fierceness of the big cat.

It was a striking statement. He had called on the *balamob*, the jaguar-protectors that had watched over his people since ancient times. He had summoned the spirit of the jaguar as a symbol of resistance. Like the jaguar, faced with diminishing habitat and the onslaught of the modern world, Dzib saw his people struggling to maintain their Indigenous identity, in the face of mounting economic and environmental pressures.

One of the direst environmental concerns in this region is the lack of water. Casandra Reyes, a biologist at the Yucatán University's Investigation Center and co-author of a paper on the potential risks of the train project, says the train will exacerbate deforestation and bring even more drought. One theory regarding the collapse of Calakmul, and many of the other cities of the Maya civilization, is that drought fueled massive civil unrest and conflict. According to some archaeologists and anthropologists, the Maya were agents of their own destruction. They squandered their resources and disrupted the weather patterns by deforesting large swaths of the jungle to build their magnificent cities. Eventually famine set in, and the peasants rebelled against the priest-kings. The bond having been broken, the peasants abandoned the great cities, spreading out across the countryside.[1]

History, according to Reyes, may be about to repeat itself. With an estimated eight thousand tourists a day arriving in the Calakmul region, all needing hotels, toilets, and showers, the obvious question is: Where will the water come from? In March 2020 a local court upheld an injunction, led by the Indigenous activist Ernesto Martínez Jiménez from Calakmul, that delayed construction of railway beds around Xpujil. Jiménez accused the National Fund for Tourism Development (Fonatur) of fraud for not adhering to the most basic international requirements for a pre-train vote meeting. Fonatur has repeatedly come under fire for not holding legitimate consultations.[2]

Fonatur, which is responsible for the Maya Train, is a powerful entity with a long and dubious past. It managed the planned development of Cancún, transforming the once-idyllic fishing town into one of the world's busiest tourist destinations. It also oversaw the Yucatán Peninsula's tourism boom, including the runaway development of the Maya Riviera, a collection of heavily developed beach towns that line the Caribbean coast from Playa del Carmen to Tulum. In January 2021, Gerardo Ceballos and more than 160 academics criticized the Tren Maya's Environmental Impact Manifestation (MIA), which the government didn't release until half a year after construction had already begun. They also accused Fonatur of deception. Instead of presenting an MIA for the construction of the entire railway, it deliberately divided the project into seven sections, ruling out the possibility of anyone being able to determine its cumulative impacts.*

Pedro Uc, a poet and land activist, is a member of the Assembly of Defenders of Maya Territory, Múuch' Xíinbal, a group of twenty-five Maya communities that have organized to oppose the train. He characterizes Tren Maya as another colonial-era swindle dreamed up by the powerful elite. "From the very start," he explains, "we said . . . that it is a project that goes against the Mayan peoples and we cannot accept something that will bring harm to our lives. . . . When you destroy territory, you destroy a way of thinking, a way of seeing, a way of life, a way of explaining the reality that is part of our identity as Mayan peoples."[3]

Among the Maya of the Yucatán, *identity* is a word frequently heard. The descendants of the Maya who fought the caste war of the Yucatán relish a redemption story about history coming full circle. They believe the snake-headed cord of life that emerged from the belly of the Maize God and connected the human world with the supernatural was deliberately severed by Spanish invaders. One day, however, they say a Maya king will assume power, and the cord will rise from the water of the Great Cenote, connecting the Maya once again to the source of life.

In the meantime, López Obrador has moved to silence dissent. In November 2021 he issued a presidential decree categorizing the train as a project of public interest and a national security necessity. His plan was to accelerate its construction in the hope that the first section of the train would begin operating before the end of his six-year presidential term. Rumor was that every few weeks, he boarded his presidential helicopter to fly over the construction site and admire what he believed would be his dazzling legacy.

---

* When his objections met with silence, Ceballos resolved to work with Fonatur as best he could. Fonatur responded by adopting a plan to build three hundred wildlife crossings that would be critical to maintaining connectivity for jaguars and other mammals. It also agreed to move the train route from the heart of Calakmul closer to Highway 186, the main east-west artery through the southern Yucatán, and to reroute the train away from core areas of existing reserves.

CHAPTER 30

A Species on the Brink

IN MAY 2021, AS CONCERNS ABOUT THE EFFECTS OF TREN Maya on jaguar habitat grew, the CITES secretary-general Ivonne Higuero announced the publication of a historic 150-page document titled *The Illegal Trade in Jaguars*. Nearly five decades after CITES, in 1975, listed jaguars as requiring the highest protection in Appendix I (see Chapter 1), its renewed attention to the species meant the illegal trade, supplying both domestic and international markets, had reached alarming levels. Higuero warned, "If unaddressed, [it] could severely impact jaguar populations and move the species closer to extinction, as it already has with other big cats, especially the tiger (*Panthera tigris*)." He instructed all jaguar range countries to "work together to . . . document, detect and deter this major threat to this iconic CITES species."[1]

The crux of the CITES report was that the survival of the species is once again threatened by a myriad of forces, including "deforestation growing inside and outside protected areas, the expansion of agriculture, cattle ranching, and human infrastructure," and the "targeted poaching of jaguars for the illegal commercial trade."[2]

For all its urgency, however, the CITES report was careful not to implicate China in the jaguar parts trade. Some conservationists suggested that CITES's reluctance was the result of its methodology. In a section titled "Information Challenges and Limitations," the report admitted that it had to contend with a limited amount of data coming from the participating countries. China, in fact, did not present any jaguar seizure data. The report also tiptoed around the subject of criminal networks, writing that the limited space devoted to them "may reflect the limitations of current evidence rather than their actual absence from the illegal trade in jaguars."[3]

John Polisar steadfastly defends the report. "I understand the heightened anxiety in the conservation community, especially after what happened to the tiger, but the report's job was to respond within the constraints of that data available. But don't underestimate its power. The CITES report represents something significant, an unparalleled buy-in among countries to counteract the forces of entropy."*

In a jaguar conservation career that has spanned thirty years, and as the former coordinator of jaguar research and conservation for the WCS, Polisar knows about responding to the data. He recently led two studies motivated by the escalation of the trade and warned that the "online trade in wildlife has expanded rapidly." He and his team called for an alliance between researchers, law enforcement, and the technology sector, recommending urgent and aggressive "electronic surveillance, undercover infiltration of criminal enterprises, capture of digital evidence, deployment of tracking devices . . . [and] the online monitoring of the wildlife trade."[4]

Although Polisar and his fellow authors were troubled by the rapid growth of online trade and the emergence of a jaguar parts pipeline between Latin America and China, they also discovered a robust domestic commerce. "I am not dismissing the fact that there is a link to China," Polisar says emphatically. "We found evidence of that. In Bolivia, it's very much the situation. . . . But what we really found

---

* In the words of Johannes Stahl, the CITES enforcement support officer who led the effort on the jaguar report, "We have to take the time to be careful and effective, not sensational." Evidence "must be bullet proof" for CITES to publish it. Juan Carlos Vasquez Murillo, chief of legal affairs and compliance, reinforced that message: "Credibility is CITES's capital."

was an unimpeded and awe-inspiring amount of domestic commerce throughout the range. That was a wake-up call."

Melissa Arias, author of the CITES report and a Ph.D. student in biology at the University of Oxford, had a similar awakening. When she first learned of China's role in the international trade of jaguar parts, she felt extreme resentment. But over the course of her research for her Ph.D. dissertation ("Illegal Jaguar Trade in Latin America"), her resentment and dismay turned to frustration. Analyzing the international media coverage, she realized that the established narrative of a vigorous illegal trade led by Chinese nationals was rife with anti-Chinese bias. While the barrage of stories helped to heighten interest in the illegal killing of jaguars, it did so, in Arias's estimation, at the expense of the jaguar by responding not to the most ecologically urgent threats but rather to the ones that attracted the most "attention, collaborations, and funding."[5]

Arias suggests that rallying around the phenomenon of the international trade, run by organized crime syndicates, made up of Chinese nationals, united conservationists across the Americas. From the media's perspective, the story generated inflammatory headlines, and from the perspective of jaguar stakeholders, the scrutiny elevated the issue into one capable of summoning CITES, the United Nations Office on Drugs and Crime, and Interpol. Once the narrative gained traction, governments and international donors took notice.

But that was only half the picture. As part of her research, Arias conducted interviews with (and distributed questionnaire surveys to) over one thousand people living in thirty-six rural villages in northwestern Bolivia in close proximity to jaguars. While she expressed empathy for them, she suggested that they were killing big cats habitually and "opportunistic[ally]" as a result of ordinary human-jaguar conflict, uncontested historic and cultural practices, ignorance of jaguar protection laws, increasing familiarity with commercial opportunities and market incentives, and a lack of jaguar-related tourism opportunities. In Belize and Guatemala, the other two countries she studied, people saw hunting jaguars as an intrinsic part of their cultural identity and often associated it with valor and "machismo."

Arias warns that because of an ecologically pure "noble savage" narrative, which perpetuates a myth of native peoples living in harmony with the natural world, the conservation community views the killing of jaguars by Indigenous hunters as defensible. But "cultural and demographic change, greater market integration, and the adoption of guns as hunting tools" has altered what were previously described as "'sustainable' indigenous wildlife use practices." Her biggest fear is that in many areas across the Jaguar Corridor where big cat populations are faltering because of the loss of habitat—and where the kinds of economic opportunities that encourage a "symbiosis between jaguars and people" that Alan Rabinowitz promoted had not caught on—the presence of poaching has already affected the species's viability.

Rabinowitz understood that throughout the Americas, Indigenous peoples engaged in jaguar hunting and trading as a cultural prerogative. But in his last years, he grew increasingly fearful that this practice had evolved into the kind of reckless trade that had already decimated the tiger population. Having reached Latin America, it had become a threat to the jaguar by creating an irresistible black market for the rural poor.[*]

In a March 2018 article titled "China's Lust for Jaguar Fangs Imperils Big Cats," Barbara Fraser investigated the emergence of trafficking routes between Latin America and China,[6] highlighting China's Belt and Road program as one of the primary culprits. She interviewed the ecologist Vincent Nijman of Oxford Brookes University in the U.K., who suggested that wildlife trafficking in Central and South America often followed Chinese construction projects.

Brook Larmer of the *Irish Examiner* holds traditional Chinese medicine responsible for much of what is happening in wildlife markets across the world.[7] The Chinese government, he writes, encour-

---

[*] He noted that traffickers were contracting with poachers and equipping them with guns and ammunition.

ages the global expansion of the $130 billion industry by lowering import taxes on wildlife items to ensure an uninterrupted supply.* Aron White, a China specialist at the Environmental Investigation Agency, backs up Larmer's assertion, claiming, "At least 46 government permits to trade in or use parts of species under the highest levels of protection have been issued to traditional medicine companies since September 2017."[8]

The Brazilian ecologist Thaís Morcatty also fears the jaguar may be next in line for exploitation. Because of the "crackdown on trafficking in tiger parts," she warns, Chinese consumers are now accepting "potentially cheaper substitutes from other big cats, including those not native to Asia."[9] In the Chinese subculture of "Wenwan," consumers are already wearing jaguar parts as symbols of status and power. The phrase Wenwan (文玩) translates literally to "toys of culture or sophistication." According to CITES, Wenwan highlights "an owner's 'taste, discernment and status,'" and its appeal comes from its speculative value, which is driven by the material it is made of. The rarer, the better.†[10]

Larmer explains that the essence of the wildlife trade is based on that very rarity. He writes that it behaves "more like drug trafficking would if opium and coca plants were in danger of going extinct," adding ominously that it is "an industry governed by scarcity" that disproportionately affects species that are already on the brink.[11]

---

* In May 2017, Beijing announced that it was developing fifty-seven traditional medicine centers in countries participating in its Belt and Road Initiative.

† Alexandra Kennaugh, a wildlife conservation and trade expert at the Oak Foundation, says the Wenwan trade is particularly troublesome because it doesn't adhere to basic "kitchen table economics." Larmer, "China's Mixed Messages."

CHAPTER 31

A Coming Crisis

ANDREA CROSTA, THE FOUNDER OF EARTH LEAGUE INTERnational (ELI), wants to make sure everyone in the jaguar world understands that their beloved big cat is next in line for exploitation. While many conservationists are unwilling to say publicly that the Chinese are driving the international trade in wildlife and the emerging trade in jaguar parts, Crosta is emphatic and unapologetic. "You can't be nonadversarial about it," he says. "The criminal exploitation of wildlife in Latin America—and Africa, too—is a billion-dollar business that's in the hands of the Chinese. The problem is that many Latin American countries can't say that, for fear of offending Chinese investors. But we know it's China. We have footage of fangs and bones from jaguars being sold as tiger parts in Guangzhou. That's the smoking gun."

Crosta first saw the heartbreaking results of the illegal wildlife trade while working in Africa as a consultant to high-tech companies producing technology for antiterrorism and homeland security. Sickened by the slaughter of fifty thousand elephants a year, he established an NGO called Elephant Action League, a site where people could anonymously report trafficking. He eventually recast it as Earth League International, what he calls the "CIA for the Earth."

Crosta has no intention of playing by what he considers the staid and conventional rules of the conservation game. "Wildlife crime is the fourth most profitable criminal enterprise in the world," he says. "It represents a convergence of multiple enterprises. Jaguar parts traffickers are also smuggling illegal timber and shark fins, drugs, even people, across borders, and they're doing it effortlessly. Same people, same network. And it can't be fought using a fifty-year-old trade agreement like CITES."

Arresting the poor ranch hand who sells a jaguar pelt and a few fangs doesn't interest Crosta. Rather, ELI's goal is to put away those who are running the trafficking rings, the kingpins. In fact, ELI's undercover agents are currently involved in seven ongoing investigations with the USFWS, the Department of Homeland Security, and the FBI.

Crosta's admirers extol his fearlessness. Jane Goodall counts herself among them: "He's passionate, he's courageous—what he's doing is very dangerous . . . and he won't ever, ever give up."[1]

The problem with cracking the illegal wildlife networks is their size and complexity. Crosta likes to tell an illustrative story of one of ELI's undercover teams in Peru being offered rhino horn from South Africa. "I could buy rhino horn in Peru," he says with astonishment. "What that means is that these networks are transnational. This is what CITES and Interpol don't understand. We need region-wide interdiction efforts, not confined by jurisdictions, and we need to be sharing information."

In 2018 ELI and two of its partners, the IUCN Netherlands and the International Fund for Animal Welfare, conducted "Operation Jaguar," the kind of unprecedented investigation into wildlife trafficking in Bolivia, Peru, Ecuador, Suriname, and Guyana that Crosta hopes will serve as a model for future operations. The investigation led to the arrest of five jaguar traffickers in Bolivia and identified seventy-five more across South America. "We were investigating twenty-five different networks," Crosta explains. "One network laundered seventy million dollars a year. Another tried to move a container of shark fins. Yet another offered three hundred jaguar fangs in the first minute of our meeting." One trafficker told an undercover ELI agent, "The Chinese mafia has hidden casinos, it runs money laundering activities and controls the cocaine business in Bolivia." In other words, operating an illegal trade in animal parts in a country that is the world's third-largest cocaine producer would be considered business as usual.

Almost two decades earlier Alan Rabinowitz discovered a similar phenomenon in Thailand's Huai Kha Khaeng Wildlife Sanctuary, where at the invitation of Thailand's Royal Forest Department, he studied clouded leopards and other wild cats. The Thais were executing big cats and laundering them across its borders in order to satisfy the booming Asian trade in animal parts. Rabinowitz never saw a single living clouded leopard, only skins. He came to understand that it was just a part of the "underbelly of a vast network of corruption" that included an active trade in nearly anything that could be bought and sold for a profit.[2] Forest guards declared war against ruthless traffickers, but the trade continued. The senior wildlife officer in the forestry department, a close friend of Rabinowitz's, ultimately was so dispirited by his inability to curb the trade that he took his own life.

Although Crosta's operations in the jaguar range have not yet produced the striking results he hopes for, Bolivia has emerged as a bright spot. In November 2021 ELI published an influential report revealing the secret details of a trafficker's business: where the jaguar parts were coming from, the trade routes, and transport methods. The report showed that Chinese-backed infrastructure projects enabled poachers to access wilderness areas. According to Crosta, the report highlighted what many people in the jaguar world dared not say: The demand for tiger parts in China, combined with the possibility of replacing them with jaguar parts, and the arrival of Chinese investments in Bolivia, aided by investor-friendly policies had created a "perfect storm" situation for jaguar trafficking.

Another enlightening report, issued not by ELI but by EcoJust and commissioned by IUCN Netherlands, asserts that conditions in Latin America are now strikingly similar to those in sub-Saharan Africa during the ivory- and rhino-horn-poaching crisis. Because of the tremendous demand in China, Asian nationals in Africa were incentivized to start smuggling operations, using the corruption and weak governmental oversight in both the origin and destination countries to their advantage. Ten years later the once-haphazard trade had transformed into an "industrial-scale" poaching operation that nearly wiped out African elephant and rhino populations.[3] The lesson for jaguar countries would seem to be that supply follows demand. Once jaguar parts are introduced as substitutes for the vanishing tiger, the jaguar could become a staple of the black-market trade.

In June 2020 the journal *Conservation Biology* published a report underscoring Crosta's argument. Compiled from data collected from all over Central and South America, it showed that seizures of jaguar parts had increased tremendously throughout the region and that private investment from China was often the instigator. According to Thais Morcatty, the report's lead author, conservationists for the first time had "a big picture of what is happening in Central and South America regarding trade in jaguar body parts." It revealed that the poaching of jaguars in South America was driven by the convergence of four conditions—local corruption, Chinese private investment, a low per capita income, and a demand for "tiger" parts in China.[4]

What is needed so desperately, according to Crosta, is increased law enforcement. Bolivia, he points out, consists of more than 424,000 square miles but has just fifty police officers dedicated to protecting wildlife. "It's like the Boy Scouts going up against Pablo Escobar," he says. "Enforcement is nearly impossible, especially when untrained officers confront sophisticated traffickers."*

Elizabeth Bennett, vice president for species conservation at the WCS, makes the point that time indeed is running out. "If we do not address this illegal trade now with local, regional and international collaboration, this could mean the end for many already endangered jaguar populations. We haven't reached a crisis point akin to what is occurring in Asia and Africa," but "we might if we don't act."[5]

---

* One Bolivian forestry police officer explained the difficulty to BBC reporters: "We can arrive on foot and [the traders] will leave in helicopters or light aircraft, or by boat down the rivers." Enever, "Bolivia Struggles."

CHAPTER 32

The Beautiful Cat

IN THE SPRING OF 2021 I TRAVELED TO THE MEXICAN state of Sonora to meet up with Alejandro Ganesh Marín, a doctoral student at University of Arizona and a *National Geographic* explorer. Ganesh ran one hundred cameras, stretching from the steep hillsides of the San Luis Mountains to the U.S.-Mexico border. It was an enormous study area that he covered almost entirely on foot. His mission was to observe how the diversity of fauna, especially carnivores, changed from one ecosystem to the next. He had videos of bears, pumas, bobcats, ocelots, white-tailed and mule deer, and even beavers. His research showed that the borderland region was a place where nature was thriving. And then one day a young male jaguar showed up on one of his camera traps. He named it El Bonito, "The Beautiful."

I started corresponding with Ganesh as soon as I heard the news of El Bonito. He told me that in early April he would be staying at the Los Ojos Ranch, owned by a Sonora-based conservation organization called Cuenca Los Ojos, where he'd be checking camera traps and setting new ones in some of the nearby washes and hills. When he said I'd be welcome to join him, I eagerly accepted the invitation.

Early one morning, just a day before I was scheduled to meet Ganesh, I drove north of Douglas, Arizona, toward the Chiricahua and Dos Cabezas Mountains. That was where the U.S. Bureau of Land Management in November 2016 had captured a male jaguar on one of its cameras, twice as far north as any jaguar had been seen in the United States for half a century. They called it a classic case of "natal dispersal," an evolutionary edict whereby young males that have not reproduced travel far from their home territories to reproduce. Some scientists and conservationists were of the opinion that pioneering males could be harbingers of a new population.[*]

I arrived back in Douglas as the sun was setting. A former boomtown that had gone to seed with the closing of a copper smelter in the 1980s, Douglas was in the midst of a recovery. A handsome old theater called the Grand was looking for $100,000 for its next phase of renovation, and the opulent Gadsden Hotel was in the process of reopening to the public. The following morning I walked across the border into Agua Prieta and waited for Ganesh as food peddlers pushed their carts up and down the dusty streets. In June 2019 the city made headlines as the scene of a Sinaloa cartel shoot-out that left twelve people dead.

Ganesh showed up in his truck, which looked like it had been used to navigate Sonora's roughest roads. He got out, grabbed my backpack, and put it in the bed with our supplies. We drove east along Highway 2, a two-lane road, scheduled to become four, that was northern Mexico's main east-west artery, through a desert of mesquite, creosote bushes, and agave. The San Luis Mountains, the northern edge of the Sierra Madres, were still half-covered in morning mist.

Our destination, Cajón Bonito, is part of the Río Yaqui watershed, which drains the rugged western and northwestern slopes of the Sierra Madre Occidental in Sonora. Like many of the tributaries of the upper Río Yaqui, Cajón Bonito flows north—within miles of the U.S.-Mexico border—then doubles back on itself, flowing south toward the Río Bavispe. Because of its south-to-north orientation,

---

[*] Melanie Culver, a geneticist and one of the heads of the University of Arizona's Wild Cat Research and Conservation Center, pointed out that research with tigers revealed that dispersing males were always the first arrivals.

Cajón Bonito was once—and is again—a bustling freeway for wildlife moving out of the heart of Sonora.

Ganesh expressed hope that El Bonito represented an emissary of a growing population of Sonoran jaguars. Gerardo Ceballos believed similarly. The famed Mexican researcher said that discovering a juvenile like El Bonito so close to the border was encouraging because it suggested that the breeding range of the species could be extending north as the big cats attempted to recolonize former territory. Though not all biologists agreed with his optimistic assessment, Ceballos was confident that El Bonito was born a mere sixty miles from the border.*

John Koprowski, Ganesh's doctoral adviser who after leaving the University of Arizona became the dean of the Haub School of Environment and Natural Resources at the University of Wyoming, was equally heartened by El Bonito's presence. "The fact that this jaguar Ganesh found is so close to the border means there are enough resources there for it to survive. The fact that we have a young male who was clearly born somewhere else shows positive signs of connectivity. Cajon Bonito serves as a functional corridor, a dispersal corridor, for jaguars and all sorts of other animals."[1]

Once it would have been considered quixotic, and unscientific, to say that animals had courage or tenacity or the ability to reason. Scientists believed they operated on pure instinct. Recently, however, zoologists have done something of an about-face regarding animal sentience, saying that more complex creatures like jaguars, wolves, and whales have highly developed personalities with characteristics akin to human traits. Some are tentative and shy, while others are adventurous. An especially intrepid animal, like El Bonito, for instance, could help a population adapt to a strange, new environment.†

---

* The breeding zone, Ceballos effused to a reporter for the *Washington Examiner*, is "now on the doorstep of the United States." Johnson, "WATCH: Experts Say Jaguar Sightings."

† According to Dr. Malcolm L. Hunter, Jr., a professor emeritus of wildlife ecology at the University of Maine, the personality trait that is common to a lot of species is the "extent to which some individuals are more curious, more exploratory." Quoted in Robbins, "Wildlife Personalities."

What all the biologists seem to agree on is that, short of reintroduction, if jaguars are to reestablish themselves in the United States, the Sonoran population needs to grow. There need to be more El Bonitos. In the last fifteen years, an assortment of people and organizations have been working toward that goal. One innovative program, initiated by the Tucson-based Northern Jaguar Project, compensates Sonoran ranchers near its 58,000-acre Northern Jaguar Reserve every time a camera on their property captures an image of a jaguar. The program, called Viviendo con Felinos, pays ranchers 5,000 pesos for every jaguar photographed on their land, 1,500 per ocelot, 1,000 for pumas, and 500 for bobcats. In the last decade Naturalia, a Mexican nonprofit, and the Northern Jaguar Project have given out almost $200,000. Because ranchers now view jaguars as an asset, many no longer shoot them or even allow the hunting of their prey.

The poaching of jaguars is still common in northern Mexico. In fact, in June 2018 someone posted online the photo of a pelt of a jaguar that had been trapped and shot. Biologists determined that it was the same jaguar that had been photographed regularly in Arizona's Huachuca Mountains in 2016 and 2017. Students at Hiaki High School in Tucson had nicknamed the jaguar Yo'oko Nahsuareo, meaning "jaguar warrior" in the Yaqui language. Yo'oko was an adolescent disperser, likely from the Northern Jaguar Preserve, and had met the unfortunate fate of other dispersers.

An hour after meeting in Agua Prieta, Ganesh turned off the highway onto a fenced dirt road, which we followed into the thorny hills. All the while, he talked about Cuenca Los Ojos and its founder, Valer Clark.

The kind of land restoration that Clark had managed over the decades—the transformation of an overgrazed and deforested landscape into an exuberant ecosystem—was nothing short of amazing. Cuenca Los Ojos's clarion call was to bring back water, soil, and life, and Clark had done just that. Where a meadow of Mexican poppies now bloomed in the late morning light, the soil was once so fragile it could not sustain grass.

Clark had come down from New York with her husband Josiah Austin on vacation in the mid-1980s, when someone suggested she buy a beat-up old ranch called the El Coronado. After the deal was done, she looked around, saw nothing but rocks and parched earth, and must have wondered what a big city girl had just gotten herself into.

Clark had grown up on the Upper East Side of Manhattan, the privileged daughter of a Wall Street broker and a high-society lady. She attended an all-girls school, and after high school and a year-long trip to Europe, she returned home and announced that she was going to be a painter. She reveled in her new identity, painted some, and cultivated a group of bohemian friends who spent their time visiting galleries and going to the theater.

Four decades later her life could not have been more different. She owned a derelict eighteen-hundred-acre ranch on the borderland between the United States and Mexico. The land was cracked and gullied and so dry, the only plants capable of surviving were creosote bushes and mesquite trees. Marshes had been drained, the area's water resources depleted by alfalfa farming, and hundreds of thousands of cattle had grazed the grass down to its roots. Drought delivered the final blow, so that when the rains finally came, water spilled out of the mountains and hillsides and washed across the desiccated land as if it were hardpan.

One day after observing how a rain shower affected a stream, Clark had an epiphany. She watched small islands of soil and vegetation accumulate behind rock barriers, and she understood that if she could harness the power of water, she could revive the land. After learning that Indigenous communities in central Mexico used rock dams to grow their crops and to control both erosion and flooding, she brought in workers to teach her how to build them and began the process of reviving the land by constructing hundreds of loose-rock dams called *trincheras*. The *trincheras*, which were as small as a few feet across or as large as a road is wide, caught and harnessed the water by slowing it down and allowing it to accumulate. On bigger washes and streams, she erected more elaborate gabions, made from riprap cages filled with rocks, sand, and soil. It took time, but pools of water and even small *ciénagas*, or wetlands, appeared, and animals that hadn't been seen in years were using the ranch again.

Clark knew that if she was going to lead a land restoration movement along the borderland, she had to convince area ranchers that what was good for the land was also in their best interests. As they saw water levels rise on their own properties and vegetation return, they began to heed her words about water conservation and soil erosion.

Inspired by her success at El Coronado, she formed Cuenca Los Ojos in 1990 to spearhead landscape-scale restoration projects. And she bought more ranches, including Rancho San Bernardino, in northern Sonora, in 1999. The ranch lay along an important wildlife corridor that connected northern Mexico with the San Bernardino Wildlife Refuge across the border in Arizona. Though early on she removed cattle from her ranches to allow native grasses to grow again, gradually she brought them back, rotating their grazing territory in an attempt to add biomass to the soil.

Eventually Clark acquired eleven properties, 121,000 acres, the majority of them in Sonora. Putting the land into a voluntary protected area program under Mexico's Natural Protected Areas system, she made certain her ranches offered essential corridors through which wildlife could migrate. And she continued to build more dams, perhaps as many as twenty thousand. The land reaped the benefits. Once Cajón Bonito had been so abjectly degraded and devoid of water that buses used the wash as a road. Today it is one of the most intact ecosystems in northern Mexico.

As Ganesh navigated the rough and winding road, we passed three cowboys on horseback. Ganesh told me that his research would reveal how cattle grazing might impact the various species inhabiting the ranch. Then he rolled down his window, which was caked in dust. He spoke in rapid-fire Spanish. I heard one of the cowboys (*vaqueros* in Spanish) say "*el tigre*" and felt an electric jolt, as I had so many times before during my travels throughout the Jaguar Corridor.

The cowboy who had spoken up smiled meekly. When Ganesh

rolled up the window, he explained that the cowboys were both fascinated and frightened by the presence of jaguars. They didn't fear the *león*, or puma, but *el tigre* was an entirely different beast. Their fear reminded me of what A. Starker Leopold had written in *Wildlife of Mexico*: "There is no animal more talked about, more romanticized, and glamorized, than *el tigre* [the jaguar]. The chesty roar of a jaguar in the night causes men to edge toward the blaze and draw *serapes* tighter. It silences the yapping dogs and starts the tethered horses milling. In announcing its mere presence in the blackness of the night, the jaguar puts the animate world on edge."[2]

About a mile down the road, Ganesh pointed to blackened trees, oaks and pines, covering the side of a hill that had burned in a 223,000-acre blaze that had scorched the borderland in May 2011. But as we dropped down into an arroyo, the color and character of the land changed. I could see the connections, the way faded foothills became arroyos, and one arroyo reached out to the next to form a series of fertile canyons.

Forty-five minutes later we began our descent into the lush Cajón Bonito with its stands of velvet ash trees, Arizona walnuts, gray oaks, hackberry, red willows, desert sumac, and bigtooth maples. As we neared the ranch house, which was set back from the river, huge Fremont cottonwoods and enormous Arizona sycamores with their bleached and bone-colored trunks lined the riverbank.

That evening, after a few hours of hiking, Ganesh and I soaked in a nearby hot spring. When I asked him how he felt the first time he saw El Bonito on his camera trap, he told me he and Kinley Ragan, a master's student, had spent a long and tiring day in the field and had just one more camera trap to check. As they were scrolling through the photos, they saw a handsome puma. A few photos later they spotted the jaguar. They lingered over the photo as if to convince themselves that what they were seeing was real. Ganesh said he had felt "a deep sense of emotion." That first photograph was revelatory, but what he appreciated even more was watching El Bonito grow. He was bulking up, as male jaguars do, his head growing bigger and boxier and his neck more bullish.

The following day Ganesh and I were out early, checking camera traps downriver in the heat-soaked hills. The first camera we inspected sat in a small tree-lined wash. Ganesh had encased the camera in protective steel because rambunctious black bears loved to swat at them.

That first camera yielded photos of a puma, two bobcats, and a spotted skunk. After refastening it, we climbed up out of the wash and onto a rise covered in cholla and prickly pear. Part of it had been torn up by rooting javelinas. The landscape appeared parched and yearning for water, but then we walked through a field of flowering pale desert-thorn and descended into an arroyo, green and oasis-like. Here Ganesh's camera had captured a bobcat, a small puma, and a cinnamon-colored black bear cub playing in a nearby puddle.

At camera number three, we hit the jackpot: a bobcat, a bear, a diminutive Coues deer, and four pumas, including a female and a cub. By the time we were bound for the fourth camera, I knew why the twenty-eight-year-old biologist had finished fifth in a national triathlon competition. I struggled to keep pace. When I told him that he navigated the landscape like a lithe and light-footed deer, he corrected me, saying that he tried to travel as a jaguar would, crossing from one wash into the next, always conscious of using his energy as efficiently as possible, and never straying too far from water.

We crossed the Agua Caliente wash and then entered the Primavera wash. Lined with large white oaks, the streambed looked like an ideal place for a jaguar. Here Ganesh lingered over a video of a huge puma with hulking shoulders and a sagging belly.

That afternoon we returned to the ranch, then drove north to the borderland. En route Ganesh broke the news to me that cartels were operating in the area, and their members were armed. They'd grown accustomed to him, but my presence might alarm them. He suggested that I leave my digital audio recorder and notebook under the seat of the truck, and he warned me not to mention anything about being a *periodista*, a journalist. If they happened to question me, I was to tell them that I was a *biólogo*, a biologist.

As it turned out, we were not stopped by cartel workers or even Homeland Security, despite the fact that we were working one hundred feet south of the international border, where the southern end of the Peloncillo Mountains straddle the line between Arizona and New Mexico. Here the Border Wall was still just a Normandy fence barricade. For the time being, jaguars like El Bonito could move between Mexico and the United States. Just west of us, however, in Guadalupe Canyon, the Trump administration had spent $450 million to build switchback roads and blast the tops off mountains in order to construct a remote 4.5-mile section of wall.

The following morning we awoke early and worked our way up Cajón Bonito, wading through the water because the banks were littered with impassably large rocks and huge cottonwood trunks that had been torn from the earth by a flood in 2019. The force of the water rushing through the canyon must once have been awesome, but now the stream, sliding through long strands of aquatic horsetail, could only be described as tranquil.

When the water deepened, I retreated to the banks, choked with thickets of birchleaf buckthorn, poison ivy, wild grapes, and climbing vines. Then I heard Ganesh call out. On the opposite side of the stream, he had found a sandbar that looked as if it had been the scene of a stampede. Tracks went in every direction—puma, bobcat, bear, javelina, turkey, coati, and deer.

We waded again into the water, moving upstream as the canyon grew narrower. Steep walls bordered the stream on both sides. Trees grew out of rock at improbable angles. Overhead two zone-tailed hawks circled.

We sloshed for another mile and then turned back, and that's when Ganesh shared his secret with me. Only months before, he had been looking at what he thought was a photo of El Bonito, when he noticed that the pattern of rosettes looked different. When he was certain it was a new jaguar, he told his advisers and named the second jaguar Valerio, in honor of Valer Clark.

As the larger, more formidable Valerio staked his claim to the area, El Bonito moved on. But El Bonito had been the pioneer, the big cat that proved Cajón Bonito was once again jaguar country.

CHAPTER 33

Jaguar Interrupted

THIRTY MILES SOUTH OF TUCSON, RANDY SERRAGLIO and I were admiring a stretch of undulating hills backed by a series of striking peaks that El Jefe had called home. Spring had settled in. The mesquite and cottonwood trees were leafing out, and a few fairy duster plants with their bright red and pink flowers bloomed on a hillside. Serraglio said in most years the hills would be an effulgent green. In southern Arizona, spring was usually the sweet spot, when the temperature was perfect, the sun shone just right, and the desert was bursting with color. But the desert country had suffered another summer of drought, followed by a winter with little snow.

As we scanned the landscape, Serraglio and I discussed El Jefe and the unhappy reality that the big cat's heartwarming story might be inimitable, at least in the Santa Ritas. El Jefe's home had become ground zero in a bellicose environmental battle over a copper mine in the middle of the mountains. The project in question was once called the Rosemont mine but now goes by the more portentous name Copper World Complex, a project the Canadian-based company says will add $700 million to the local economy over twenty years.

As Russ McSpadden of the Center for Biological Diversity wrote in the *Arizona Daily Star* in a poignant and personal so-long to the Santa Ritas, our appetite for copper is growing. We demand "tons of copper rolled into our internet lines, smartphones and electric-vehicle batteries."[1] If we are to become copper self-sufficient, we will need to extract it from places that are precious to us.

The Santa Ritas contain enough copper to account for roughly 10 percent of U.S. production. McSpadden wrote that if mining went forward, "Several mountain peaks . . . are scheduled to die," because the "Toronto Canadian mining giant Hudbay Minerals is determined to decapitate them."

The sheer massiveness of what Hudbay is proposing is hard to fathom: a four-decade project, connecting Copper World on the west side and the Rosemont site on the east, that would require a 1.5-mile stretch of scenic ridgeline to be removed.

Currently, Hudbay's sights are focused on the western side, on 4,500 acres of private land that it says contain sufficient copper to support a fifteen-year mine life. That mine alone would create 64 million tons of mine waste and would necessitate pits and piles for tailings and waste rock.

Another concern is the mine's impact on hundreds of springs and over 150 ephemeral streams on which wildlife and possibly future jaguars would rely.* Those springs and streams were on Serraglio's mind when he drew my attention to a fertile line of tall cottonwoods to the east. "That's Cienega Creek," he said. "It's not just life support for twelve endangered species; it provides twenty percent of Tucson's natural groundwater recharge. The mine will suck it right down. Instead of water, Tucson will get mining dust."

AFTER A SHORT HIKE, Serraglio and I drove east into the Canelo "Cinnamon" Hills, crossed the Huachucas, and headed south to see the Border Wall.

---

* Nearly 80 percent of Arizona lacks any semblance of groundwater regulation, and where the water is protected, mines are often exempt from following those rules.

Nearing the wall, we parked on the side of a sandy road and walked south at the edge of Glenn Spencer's land. Spencer is the president and founder of the American Border Patrol, a group dedicated to establishing a "shadow Border Patrol" by using citizen patrols and sensors and surveillance equipment to track the movement of migrants crossing the border. Spencer and his ilk believe there is a secret Mexican conspiracy to reconquer—*reconquistar*—the U.S. Southwest, and that migrants are the country's forward presence.

We skirted the eastern border of Spencer's property, which was decorated with American flags. Along the fence line, Spencer had placed a skeleton in a lawn chair. The skeleton smoked a cigar, wore a MAGA hat, and held a Bud Light in one of its hands. It faced south, and with its other hand, it flipped off Mexico.

Serraglio and I approached the wall, which loomed over the landscape like an ancient beast, an edifice to a cry conjured on the campaign trail: Build That Wall. That "big, beautiful Wall."

I'd read about the Wall's dimensions, but until I stood next to it, the numbers had no meaning. The Wall, capital W, which replaced a shorter wall and head-high vehicle barriers, is made up of six-inch-wide steel poles known as bollards. The construction crews sunk the thirty-foot-high bollards deep into the ground, filled them with rebar and concrete, topped them with anticlimbing plates, and strategically separated one from the next by three inches of space, preventing migrants from squeezing through but allowing Border Patrol agents to monitor activity on the Mexican side.

Adjacent to the Wall, they carved out a 150-foot-wide scar of an "enforcement zone" that includes an access road used by Border Patrol vehicles. The next step is for the crews to add ground sensors, lighting, and IFTs (integrated fixed towers), as they have in other areas along the wall, with audio and video capabilities that can capture the sight and sound of a Gila monster scraping across the sand at dusk. One day the IFTs will be incorporated into a vast system called Integrated Surveillance Towers, forming a virtual wall that a decade from now, when fully operational, will render the physical Wall redundant.*

The Wall, which Donald Trump made a symbol of his administration's efforts to halt illegal immigration—a complex and grave issue that has vexed many administrations—was built quickly, and expensively. Its $15 billion price tag (Trump never got Mexico to pay for it) included 800,000 tons of concrete and 600,000 tons of steel. The money for it had to be appropriated from the Defense Department. From an environmental perspective, the project stirred controversy from the start. Big federal construction projects normally require an environmental impact assessment, but the Trump administration waived countless laws along the 1,954-mile border to speed up construction, claiming that erecting the Wall without delay was essential to national security.†

The wall crossing the nearby San Pedro River is especially expensive. In fact, it's the costliest section in all of Arizona. Environmentally, it may also be the most destructive. The San Pedro is one of the richest riparian areas in all of North America. Four hundred species of birds inhabit the immediate river area or migrate through using the corridor. Birders from all over the world come to the San Pedro to see the brilliantly colored and elusive eared quetzal or the wary Abert's towhee or the valley's fifteen kinds of hummingbirds. The San Pedro's abundance also includes eighty species of mammals, forty species of reptiles and amphibians, and fifty-five rare or endangered species.

The last major free-flowing river in the state of Arizona, the San Pedro, looks underwhelming. When I finally stood along its banks,

---

* In 2005 Congress passed the Real ID Act, which allowed the Department of Homeland Security to seize any and all land along the southern border, including confiscating private property through eminent domain. In 2006 the Secure Fence Act directed its secretary "to take appropriate actions to achieve operational control over U.S. international land and maritime borders."

† Warner Glenn, whose Malpai Ranch runs along the border, opposes the Wall. "You're talking seventeen million dollars a mile to put something like this in," he said. "But as far as terrorism or drugs go, it's a joke. They're going to get it in some way. So the Wall is not worth the cost."

I saw an unassuming ribbon of green, populated by Goodding willows and substantial Fremont cottonwoods that were shedding their first seeds, but only a trickle of current, more a riverbed than a river, no deeper than the water in a backyard birdbath and no wider than a sidewalk.

The San Pedro begins in Mexico and flows north into the United States. On a map, it carves a fairly straight south-north line, beginning as mountain runoff in the Sierra Madres and following a broad valley into southern Arizona's Sky Island region. From there, it flows between a series of obscure and arid mountains. On the ground, however, the San Pedro meanders through *ciénagas* (natural bogs), thick mesquite bosques, and densely treed riverbanks. Not so long ago Congress had the foresight to make the San Pedro River the nation's first Riparian National Conservation Area, protecting some 56,000 acres and forty-seven miles of river habitat between the border and St. David, Arizona.

The faint flow of water I saw just north of the Wall was, sadly, indicative of what is now happening to many other parts of the San Pedro. There are dry segments where the river no longer reaches the surface. Climate change, groundwater pumping for industry and for drinking water for Tucson's expanding suburbs, and the mixing of tons of concrete for the Border Wall are the culprits, depriving the San Pedro of ancient flow. Monsoon rains once inundated the region from June through September, but they have become far less reliable, often leaving groundwater alone to sustain the river. But when the rains do arrive, the serene San Pedro barrels north, roaring like a freight train. Sometimes it gets so much rain that water levels reach an unimaginable seventeen feet.

How the Wall will change the river's flow, no one quite knows. Initially, the U.S. Border Patrol rejected the design, claiming it ignored the river's replenishing floods. In June 2020, however, officials unveiled a new plan that incorporated a series of gates that would allow floodwater and debris to pass through. While environmentalists applauded the change, they worried that engineers stopped short of designing what the river really needs. The gates, they pointed out, are not hydraulic and will require border agents to open and close them manually. Many wonder how quickly they will be able to respond when a fierce summer storm sends floodwater crashing up against the barrier.

These same people would suggest that the gates are symptomatic of an overall problem—rushed construction. A case in point is the Wall's eight-by-eleven-inch "cat passages," which have been widely ridiculed. An op-ed cartoon in the *Arizona Daily Star* by Rob Peters, a biologist and executive director of Save the Scenic Santa Ritas, showed a bewildered Sonoran jaguar stuck helplessly in a cat gate, his front half on one side of the border, and his back half on the other. The jaguar says, "I knew I shouldn't have eaten that last armadillo," to which a coati responds, "Welcome to the US, O Great Spotted One." Meanwhile an onlooking frog proposes using Crisco.[2]

According to a group of scientists who co-authored a paper in *BioScience* titled "Nature Divided, Scientists United: US–Mexico Border Wall Threatens Biodiversity and Binational Conservation," the cartoon captures the general feeling in the conservation community. The Wall went up so fast that "all the mitigation and study measures" normally required "never got done."[3] The long-term tragedy, according to Aaron Flesch, one of the paper's authors, is that the Wall will disrupt the "connectivity of plant and animal habitats" and will undo "more than a century of binational investment in conservation." By literally dividing a continent, it will put an end to the historic flow of species and will almost certainly terminate the dreams of jaguar biologists who hoped that Mexican jaguars like El Jefe, wandering the borderlands, would somehow be able to resuscitate the U.S. jaguar population.*

The biologist Chris Bugbee, who with his loyal dog Mayke had tracked El Jefe for more than three years, says that the Wall will end the story of migration. Absent the border wall, he believes it would be just a matter of time before jaguars, already adjusting their range

---

* More than 2,500 scientists endorsed the article, over two thousand of them from the United States and Mexico.

father north because of climate change, would cross the border, and eventually a female would arrive in Arizona or New Mexico.

Turning our backs to the Wall, Serraglio and I listened for the sheltered songs of birds high up in the cottonwoods. Having heard the sweet chirping of a yellow warbler, one of the area's first migrators, Serraglio was scanning the canopy of tree limbs with his binoculars. Eventually, we left the river and headed back to the truck, and our conversation returned to El Jefe. "He gave people hope," Serraglio said. "He allowed them to believe there was still wildness left in the world."

CHAPTER 34

Bringing Jaguars Home

KEENLY AWARE OF THE ONGOING DISCUSSION IN THE JAGuar community regarding the viability of U.S. jaguars, I set out on a backpacking trip on the Arizona Trail to explore the U.S. region of the Borderlands Secondary (Jaguar) Area. My plan was to cross the Huachucas, the Canelo Hills, the Patagonias, and the Santa Ritas, then jump off somewhere outside Tucson.

Prior to my hike, I'd talked with Matt Nelson, the director of the Arizona Trail Association. Nelson said he'd envisioned the eight-hundred-mile route not only as a recreational trail but also as a "big-landscape wildlife corridor." He explained that if jaguars were ever to recover their historic territory in the Southwest, which reached clear to the southern rim of the Grand Canyon, they would likely move through the state via the Arizona Trail.

My hike began 1.8 miles north of the U.S.-Mexico border and the Arizona Trail's southern terminus, the Montezuma Pass Trailhead at 6,569 feet. Standing in the parking lot, my hiking partners and I listened to Randy Serraglio give an impromptu talk about El Jefe to a spellbound group of birders.

When Serraglio finished his story, a woman asked if a jaguar in Arizona would constitute an invasive species. She expressed surprise that an animal she associated with steamy jungles could live—or would choose to live—in a thirsty land where barely a dozen inches of rain fall each year. Having heard the question before, Serraglio chuckled and told her and the other doubters that he had four resounding words for them. "*This is jaguar country,*" he said categorically. Then he added, "This has *always* been jaguar country." Jaguars are what biologists call "generalists," he explained, consummate adapters that can habituate to all kinds of different habitats, from the rainforest of Belize to the mesquite-and-juniper-covered rises of the Canelo Hills. But then he cut his speech short, as we needed to climb over 2,500 feet in the next four miles under a blazing sun.

By the time we hit the Miller Peak Trail Junction, the spur trail to the highest point in the Huachucas, my pack felt like a dead javelina on my back. The ascent had been strenuous, and I was grateful for the brief stop when Serraglio showed me a camera trap, anchored to a ponderosa pine, that had photographed a jaguar named O:ṣhad Ñu:kudam ("Jaguar Protector" in the Tohono O'odham language) just days before we began our hike.

For the last mile and a half to Bathtub Spring, we encountered long stretches of ankle-deep mud and spring snow. Alongside the trail, I noticed a discarded Electrolit bottle and a serape that had undoubtedly once belonged to an immigrant so desperate to escape the Border Patrol that they had chosen to pursue freedom in the United States via the rugged Huachucas, where nighttime temperatures hovered near freezing and the days sizzled with heat.

It was a cool night and an even cooler morning. At five a.m. I could see my breath. By eight a.m., Serraglio and our friend Nathan, who had carved out climbing routes in the peaks of southern Chile but had never been to Arizona's Sky Islands, were headed back to Montezuma Pass, while my longtime buddy, the writer David Gessner, and I pressed on.

The thing that caught my eye after we crested a ridge and began a long descent into a green and fertile canyon was how beautifully deceptive the Sky Islands are. From a distance they look sun-beaten and inhospitable, arid, desiccated even. But once you get back into the mountains, they are dotted with thriving climates and watery oases that teem with biodiversity. Especially in April, they are full of lush creek beds, drainages running with snowmelt, and birds, from diminutive Arizona woodpeckers to summer tanagers capturing insects in the still air. An ecoregion that stretches from western New Mexico and southeastern Arizona south to the northern Mexican states of Chihuahua and Sonora, the Madrean Sky Islands are named for a series of mountain ranges isolated from each other by miles of rolling and open grasslands, desert, and now expanding housing developments. The Sky Islands connect the wilderness of the Rocky Mountains with Mexico's Sierra Madres, and that incomparable interaction between north and south produces a striking biodiversity. In the hills and arroyos, ash, pine, and red willow mingle with mesquite, prickly pear, and cholla. And jaguars drink from the same watering holes as black bears, bobcats, and Mexican gray wolves.

That second evening Gessner and I bivouacked early at Rattlesnake Spring. We talked about CANRA, the newest addition to the discussion of U.S. jaguars. The Central Arizona–New Mexico Recovery Area is based on a proposal by a multidisciplinary team of authors, many of whom have devoted their lives to studying jaguars. Using evidence assembled by WCS researchers who spent the last decade compiling 350-plus recorded jaguar sightings in North America, dating back nearly two centuries, the team published their findings in the journals *Oryx* and *Conservation Science and Practice*.[1] Their bold assertion was that CANRA had the potential to support a self-sustaining population of 90 to 151 jaguars that would be viable for at least a century.

At 20 million acres, CANRA would be as large as the state of South Carolina. It would extend across the Mogollon Plateau, an elevated and remote landscape stretching from the Gila National Forest in New Mexico to the Grand Canyon, where despite the two-decade drought in the Southwest, water is plentiful.

On its east end, CANRA would be anchored in the Gila and Aldo Leopold wildernesses, forming a block of wild, roadless coun-

ing three-quarters of a million acres. On its west end, it runs from the Superstition Mountains east of Phoenix north to the San Francisco Peaks outside Flagstaff. Key to CANRA's potential success is its relative emptiness and isolation; only 11 percent of the land is private. The bulk of it is owned by the federal government, while tribal sovereign nations (the White Mountain and San Carlos Apache) and state agencies manage the rest.

For Eric Sanderson and Kim Fisher, this newest effort is a particularly personal one. In 2018 the Jaguar Recovery Team (JRT), of which they were part, released its recovery plan (though it would not release it publicly until April 2019), designating 56,049,920 acres of critical habitat. The JRT called this habitat the Northwestern [Jaguar] Recovery Unit (NRU) and anticipated that the region's carrying capacity could exceed 3,400 jaguars.

The team divided the NRU into two core areas and two secondary areas. It then divided one of those secondary areas, the Borderlands Secondary Area, into two regions—Mexico and the United States. The U.S. region encompassed 764,207 acres.* The JRT established the northern edge of the jaguar range sixty-eight miles north of the U.S.-Mexico border to Interstate 10, the southernmost U.S. transcontinental highway. Its maps allowed for a total of forty-two jaguars in the entire Borderlands Secondary Area, with a mere six jaguars inhabiting the United States.

Sanderson and Fisher felt that the JRT had based its decision on a narrow-minded view of jaguar habitation. They argued that the ESA required the USFWS to establish its recovery plans on the big cat's historical range and pointed out that Arizona was once a jaguar stronghold, with jaguars inhabiting the northern half of the state, the rugged Chiricahua Mountains, near the border with New Mexico, and the Coronado National Forest units north of Interstate 10. Others accused the JRT of deliberately selecting a huge recovery zone. If jaguars were to recover anywhere within the vast Northwestern

---

* The U.S. region extended from the San Luis Mountains in New Mexico to the Baboquivari Mountains in Arizona.

Recovery Unit, the USFWS could declare success, without a single big cat inhabiting the United States.[2]

Now, as part of the new team of authors, Sanderson and Fisher advocated not just for an expansion of critical habitat but for an even more dramatic commitment to restoring jaguars in the United States—reintroduction.*

The proposal to establish "an experimental population under authority of section 10(j) of the Endangered Species Act" in New Mexico's Gila National Forest met with predictable acrimony, even though the authors suggested that the population be designated as "nonessential."[3] A nonessential population designation relaxes ESA protections and gives wildlife managers considerably more freedom and flexibility, including the option of removing individuals that prey on domestic livestock.

In 2021, however, New Mexico livestock groups took aggressive measures to undo the results of the jaguar recovery plan by filing a successful lawsuit prompting the removal of all the proposed acreage in that state—nearly sixty thousand acres. Chad Smith, CEO of the New Mexico Farm and Livestock Bureau, who celebrated the verdict, said, "To think of a jaguar and its opportunity to thrive in the arid southwest is absurd."[4]

Michael Robinson and the Center for Biological Diversity think otherwise. In late 2022, Robinson sent a letter and a petition to secretary of the interior Deb Haaland (who has long supported imperiled species) and Martha Williams, the head of the USFWS.† As the petition's lead author, Robinson made a compelling argument for expanding critical habitat lands in the United States and for the experimental—and controversial—introduction of a breeding pop-

---

* Regarding the reintroduction of jaguars, the group of authors contended that given the nearly impermeable Border Wall, expanded critical habitat designations would not be enough to ensure that a breeding population of jaguars would be capable of gaining a foothold in the United States.

† As a top attorney for Montana's Fish, Wildlife, and Parks and later its director, Williams had earned respect from both sides of the political fence during the bitter wolf wars that engulfed the state.

ulation of jaguars in the Gila National Forest, citing the USFWS's own research, which concluded that the chance that jaguars would naturally settle the Sky Islands was "very small, even under more optimistic" circumstances.*⁵

Jim Heffelfinger, wildlife science coordinator for the AZGFD, called the plan a political powder keg. "We don't need this kind of negative uproar about wildlife when we could focus our conservation efforts on where jaguars want to live," he said.⁶ "All they [Sanderson and his team] really did was to take the USFWS habitat suitability model and increase the elevation for suitable habitat from 2000 meters to 2400 meters. We get caught up with sexy spotted cats, but with so many species currently listed under the ESA needing our help, how important is it that we pour millions of dollars of ESA time and money into recovering the jaguar?"†

Heffelfinger's colleague, Clay Crowder, who acts as assistant director of the Wildlife Management Division and oversees the AZDGF's much-contested Mexican wolf program, responded similarly. "It's a matter of department priorities. We're at the extreme northern periphery of the jaguar's habitat. But I understand it. People love charismatic species. They love the idea of a lone jaguar wandering the mountains. But would they tolerate dozens of them?"

USFWS biologist Marit Alanen says, "When thinking realistically about funding for recovery projects, it's important to consider how expensive any sort of captive breeding and translocation process would be."⁷

---

* In a study in 2013 (prior to the expansion of the Wall), the biologist and Northern Jaguar Project board member Peter Warshall predicted, based on "dispersal information on female jaguars," that it would take "anywhere from 44 to 243 years for the first female jaguar to naturally appear in the Sky Island mountains of southeastern Arizona or southwestern New Mexico." Warshall, "When Will Female Jaguars."

† Heffelfinger points out that a population viability analysis done by Phil Miller, director of science at the Conservation Planning Specialist Group and co-author of *Guidelines for Species Conservation Planning* (published by the IUCN), concluded that we would need to translocate thirty to forty jaguars to establish a genetically viable population, which is "double the number of big cats currently living in northern Sonora from where the jaguars would likely be sourced."

Based on the principle of "trophic rewilding," an ecological restoration strategy that uses species introductions to restore natural and historic interactions and self-regulating ecosystems, the prospect of jaguar reintroduction has gained an air of respectability, despite current state and federal opposition.*

Perhaps the world's most ambitious rewilding experiment—and a possible model for CANRA—is taking place in the expansive wetlands ecosystem known as the Iberá (in Guaraní language, *y berá* means "shining waters"), a region in northeastern Argentina bordered by Paraguay, Uruguay, and Brazil. There in the 1990s Doug and Kristine Tompkins purchased over 370,000 acres of degraded ranch land. Less than a decade later, via their Conservation Land Trust, they helped to establish a rewilding program, run by their in-country partner Fundación Rewilding Argentina, to reintroduce native species like river otters, tapirs, peccaries, giant anteaters, and jaguars to the ecosystem. In 2015 Doug Tompkins died tragically from severe hypothermia after his kayak capsized in forty-degree waters; just days later Kristine Tompkins donated the entirety of their landholdings in the Iberá to the Argentine government with the strict stipulation that the land be part of a national park. In 2018 the 1.75-million-acre Gran Iberá National Park became a reality.

Three years later conservationists from Fundación Rewilding Argentina opened up a pen that held two four-month-old jaguar cubs, Karai and Porá, and their mother, Mariua, and for the first time in the history of the species reintroduced jaguars into the wild. When the big cats flourished, Fundación Rewilding Argentina released seven more in an attempt to expand the population's genetic diversity, and in July 2022 two cubs were born from that reintroduced population.

"The situation of the jaguar in Argentina is so critical," Sebastián Di Martino, the conservation director of Rewilding Argentina,

---

* Sergio Avila, a biologist for the Sierra Club, points out that even if it were politically possible, translocating jaguars to the CANRA recovery unit would be problematic. Jaguars from Arizona and Sonora are part of an edge population that possesses genetic adaptations to hot, arid conditions that jaguars captured from other parts of the range and translocated to CANRA lack.

explained. "I think one of the main messages here is that we need to do these kinds of things—things that maybe ten years ago were considered to be too risky or too crazy or too audacious."[8]

IN THE TOWN OF PATAGONIA, my buddy David Gessner left for Tucson, and Randy Serraglio rejoined me for the last leg of the hike through the Santa Ritas. Just two days in I could see why El Jefe had elected to spend so much time in these rough and secluded mountains and canyons. Serraglio and I, taken with the beauty of the terrain, made a wrong turn and traipsed far up into a gulch before we realized we had strayed. At one point, we came upon what Serraglio identified as a scrape, a symmetrical collection of leaves that he thought might have been made by a jaguar marking his territory.

For the rest of the day, the forest held a wilder kind of energy for us, the same kind of mysterious vitality that had so captivated Aldo Leopold a century ago. In "The Green Lagoons," Leopold wrote of a three-week paddling adventure he and his brother Carl took in 1922 to the Colorado River Delta. Fascinated by the wild landscape where jaguars roamed, he wrote, "We always examined these deer trails, hoping to find signs of the Despot of the delta, the great jaguar, *el tigre*. We saw neither hide nor hair of him, but his personality pervaded the wilderness. No living beast forgot his potential presence, for the price of unwariness was death. . . . No campfire died without talk of him."[9]

By the time we discovered our mistake, it was too late in the day to try to make our way back. The light, we knew, would fade more quickly in the steep-sided gulch. So as the day's breeze waned, we pitched our tents under a canopy of tall and still ponderosa pines. Just as we were bedding down for the night, we heard the creaky calls of birds in the branches of a tree above our heads. It sounded like kids blowing on thick blades of quack grass poised upright between their thumbs. The birds called for a minute or two, then moved on, and when they did, Serraglio exclaimed from his tent, "Those were Mexican spotted owls! A family of them." The owls were the shyest of creatures, he said, and in over twenty-five years in the field as an avid birder, he'd heard them only a few times before. Not long afterward the night squeezed all light from the sky, and darkness pressed in, tight and coal black. Half an hour later I watched through the roof of my tent as the vivid white bands of the Milky Way came to life.

The following morning we woke early, tired but not particularly troubled by our inadvertent detour. We continued our hike through the Santa Ritas and even camped at Cave Creek, where Chris Bugbee's camera had captured the footage of El Jefe that made its way around the world.

Today many conservationists say that any hope jaguars will repopulate the desert Southwest on their own—even as climate change pushes them north—has pretty much been extinguished by the presence of the Wall. Jaguars like El Jefe once hopscotched across the border by way of the Sky Islands, where they could travel safely and hidden from sight. They can still cross by way of Arizona's unobstructed washes. But for all intents and purposes, the state's 225 miles of Wall will likely stop them in their tracks. To quote Myles Traphagen, a conservation scientist with Wildlands Network, the Wall will "alter the evolutionary history of North America forever."[10]

The probability that jaguars will not be able to traverse the Wall gives advocates of the CANRA proposal and Michael Robinson their heightened sense of purpose. If the Center for Biological Diversity's petition is met with enthusiasm, the dream of Eric Sanderson, Kim Fisher, Tony Povilitis and all the others of returning a jaguar population of consequence to the American Southwest might become a reality.

Writing plaintively, Robinson portrays reintroduction as an essential act of atonement. "It is good for the national soul," he writes, "to take responsibility and make right the mistakes made by the nation in the past. . . . In this instance, it is not too late to fix what was thoughtlessly broken. Let us . . . welcome them home."*[11]

---

* Michael Robinson points out that producing the petition took a "village." Though his name is attached to the document, over twenty people contributed to it.

Epilogue

THE WEATHER WAS WITHERINGLY HOT. IT WAS THE KIND of steamy day when even the birds, huddling in the canopy of the trees to escape the sun, seemed unable to summon the energy to call out.

Four Maya women, adorned in colorful dresses, danced on the veranda, while Maya men played marimbas, striking welcoming notes with their mallets. People filed in and took their seats under the tent. They were there to honor Alan Rabinowitz, to remember the life of a legend, and to participate in the dedication ceremony of the Dr. Alan Rabinowitz Research Center.

Five years after his cancer diagnosis, Rabinowitz established the Jaguar Corridor, one of the most ambitious conservation projects ever undertaken. Then he inspired reluctant Latin American governments to consider adjusting their development plans to support and protect it. When he finished, he embarked on a challenging new project, the Journey of the Jaguar, an adventure in storytelling that was supposed to take Howard Quigley and him across the corridor, from Arizona to northern Argentina. His intention was to make brief, captivating videos for social media along the way. That dream was cut short. But not long before his death, he traveled to Colombia's Darién Gap, hoping to save a forest and its jaguars from miners, prospectors, and poachers. He also contacted Eduardo Carrillo about visiting Costa Rica's Guanacaste National Park, which has the highest concentration of jaguars in Central America.

Alan Rabinowitz's family, his wife Salisa, and his children Alexander and Alana had been organizing the memorial for over a year.

The speakers, Salisa, Alexander, and Alana, and longtime friends, Andy Sabin, Greg Manocherian, and Michael Cline, who along with Tom Kaplan provided the majority of the funding for the Dr. Alan Rabinowitz Research Center, spoke glowingly of the man they loved and admired. In an emotional speech, Bart Harmsen, the Panthera Belize director, talked about reading *Jaguar* for the first time in 1994 as a student in his home city of Amsterdam. "The man touched me," he said, "like he touched many other people." Harmsen put his studies on hold and bought a ticket to Belize. He came with a backpack and his copy of the book. The only thing he knew was that he wanted to study jaguars. He wrote to Rabinowitz who, remarkably, responded. "Alan always wrote back," he said, "to young people—he did that—because that was important to him." Eventually, Harmsen got a grant to do his Ph.D. on jaguars in Belize under Rabinowitz's watchful and demanding eye. For five years he lived in Rabinowitz's shack and embarked on his own adventure. But it was Rabinowitz's original story that inspired him. Without that story, he would have remained in Amsterdam.

After the eulogies, the family lovingly dispersed Rabinowitz's ashes and those of Howard Quigley, his lifelong friend. Quigley had died just months after our trip to the Yucatán. He had been diagnosed with metastatic prostate cancer and had been participating in a medical trial, to which he had responded well. But in July 2022 his health took a dramatic turn, and he died quickly, just two months later. His death shook the world of jaguar conservation, where he was regarded as an esteemed and groundbreaking biologist as well as a profoundly decent man.

A young Alan Rabinowitz had helped to save the Cockscomb Basin and its jaguars and helped to kickstart a conservation movement in Latin America dedicated to a misunderstood predator. Thanks to Rabinowitz, not too far from Cockscomb, people and jaguars are now living in close proximity to each other in a highly altered landscape, and both are thriving. Elsewhere along the range, after hundreds of years of bitter enmity, cattle ranchers and jaguars, sharing the same land, are learning to exist side by side.

Perhaps the most far-reaching and promising of all the current jaguar conservation efforts is the 2030 Roadmap, a spectacular plan spearheaded by Panthera, the WCS, the WWF, the United Nations Development Program, United Nations Office on Drugs and Crime, United Nations Environment Program, CITES, and the Convention on Migratory Species, to connect key jaguar conservation areas by the year 2030 and to secure thirty high-priority jaguar landscapes. By rejecting the idea that merely defending and maintaining vital ground will protect the species from extinction, the Roadmap, to which sixteen of the eighteen jaguar range countries have committed themselves (Venezuela and Guyana being the only holdouts) is the most significant, range-wide effort to ensure the future of the species since Alan Rabinowitz's Jaguar Corridor Initiative. Importantly, it also takes aim at jaguar poaching and trade, seeking to implement aggressive enforcement controls, cross-border cooperation, and strict punishments for offenders.

Apart from the 2030 Roadmap and the illegal trade in parts, no issue in the world of jaguar conservation is as topical as U.S. jaguars. Today two male jaguars roam the mountains of southern Arizona: Sombra (Spanish for "shadow"), one named by students at Paulo Freire Freedom School in Tucson that has called the Chiricahua Mountains home since late 2016; and the newest addition, a jaguar named Cochise.

In late 2023 Jason Miller, a wildlife videographer, posted a nearly five-minute video on YouTube that he subtitled "An Arizona Trail Cam Journey."[1] He'd been running wildlife trail cameras for five years, hoping that one day he'd capture footage of a jaguar. Smiling broadly, he announced that at 8:30 p.m. on December 20, 2023, he'd succeeded. Appearing on camera, Miller introduces the next seventeen-second clip, saying, "Unbelievable, here it is." What viewers see is a close-up of a jaguar sniffing the ground. With crickets trilling, it lifts its head and opens its mouth wide, as if showing off its canines. Its eyes glow in the night. "What a magnificent creature," Miller utters. "If unnamed, I will name it Cochise." The video quickly collected over 300,000 views and almost two thousand comments.

To keep the jaguar's whereabouts a secret, Miller offered no details other than the fact that he had captured the footage north of the Mexican border, in "a deep canyon." In early January 2024 the AZGFD verified the authenticity of the footage, adding that Miller's discovery marked the eighth individual jaguar—all of them males—spotted in the wild in the United States since the 1990s.

The video elicited impassioned calls for the reestablishment of jaguars in the Southwest. But just one month after Miller's electrifying footage appeared on YouTube, the movement to bring a breeding population of jaguars back to the United States suffered a setback, when the USFWS rejected the Center for Biological Diversity's petition for the experimental introduction of jaguars.

Tom Paterson, president of the New Mexico Cattle Growers Association, breathed a sigh of relief, as did many other cattle ranchers in the Southwest. Some of their concerns were justifiable. But Paterson immediately relied on an unscientific and risible portrait of the jaguar. "The jaguar," he said, "would be killing our elk. It would be killing our recovering deer herd, and it would be killing our cattle and our horses. They would be potentially terrorizing those members of our rural communities."

Many hunters, who are still bitter over the reintroduction of the Mexican wolf twenty-five years ago, were also relieved. Don McDowell, Sr., president of the Arizona Deer Association, in an interview with *MeatEater*, was emphatic about his organization's objections. "We're diametrically opposed to the reintroduction of jaguars," he said. "The jaguars we have here are rogue. None of their historic range is in Arizona."[2]

Jim Heffelfinger, who, in addition to working for the AZDGF, is the chairman of the Western Association of Fish and Wildlife Agencies' Mule Deer Working Group and author of the authoritative *Deer of the Southwest*, was more measured. "My gut feeling," he says, "is that they wouldn't have the impact of wolves. I don't have a fear that they'll devastate wildlife. But my short answer is, I don't know."

John Polisar, one of the authors of the proposal to bring back jaguars, discounts the notion of them decimating cervid herds. "I understand their concern," he says, "but it doesn't worry me too much." Jaguars, he points out, have low population densities—especially in a desert landscape—and low reproductive capacity. He stresses, too, that jaguars are not exotics. They are part of the native fauna; they belong on the landscape.

Meanwhile Michael Robinson, a hardened veteran of conservation battles, took the USFWS's decision in stride, adding, "We're very disappointed, but this is by no means the end of the road for our advocacy."[3]

Not all jaguar biologists believe that we should reintroduce big cats into the desert Southwest. Alan Rabinowitz championed natural dispersal. Allison Devlin and others at Panthera agree that even with the Wall natural reestablishment offers the best opportunity for success. "Give jaguars space, protection, connectivity, habitat and a healthy prey base," Devlin says, "and they will eventually roam once more in their historical range. However, this will be a challenge for the species going forward. It all begins with human-cat coexistence."[4]

Everyone in the jaguar world can agree that across the jaguar corridor, big cats are facing dramatic challenges. In just the past fifty years, over half their historic range has disappeared. Apart from the jaguar success stories in places like the Pantanal, real victories have been hard to come by. But the trends of loss, according to John Polisar, can be reversed. "We have to maintain the connective tissue of the corridor, which requires making more sections legally defensible, and protect the stepping stones, the JCUs. Some have shrunk over time, while others have suffered from fragmentation."

Then there is the jaguar itself, which despite innumerable obstacles, persists in finding its way across treacherous landscapes, drastically altered by the modern world, more often than not managing to evade the terrible bullet.

Like her mentor Rabinowitz, Allison Devlin continues to harbor hope for the species, and it is her unwavering belief in its adaptive abilities that she clings to. "I remain cautiously optimistic," she says. "These are resilient, wide-ranging animals. They see the landscape differently than humans do and given the room to roam they can succeed."[5]

And that is the key to the jaguar's future—the resourcefulness that allowed it to cross the Bering Land Bridge more than a million years ago and eventually populate five thousand latitudinal miles of the New World. The landscape it encounters today is even more perilous than the New World once was. But as Alan Rabinowitz noted, the jaguar is a fighter; it has the "audacity to survive."[6]

Acknowledgments

FIRST ON MY LIST OF PEOPLE TO THANK IS MY MOTHER, Renate, who has never wavered in her belief in me. A voracious reader and a lover of life, even at ninety-three she makes our days better with her warmth and affection, her laughter, and her adventurous spirit. I'd like also to thank my daughters, Aidan, Rachel, and Willa, who, while off exploring the world and pursuing their own passions, fill my life with joy. Thanks, too, to the extended Campbell clan, who remind me of the gifts of family. And to my faithful buddies who teased me endlessly about going MIA for years but never stopped inviting me to join them on backpacking, fishing, and hunting adventures.

Every book is a collaboration, and I owe an enormous debt of gratitude to my longtime agent David McCormick, who helped me transform a passion project into a marketable book and gamely read my sprawling rough draft, assisting me in turning it into something readable. I am deeply grateful to my insightful editor John Glusman, an unflappable forty-year veteran of the publishing industry whose talents, faith, and encouragement made this book so much better than it might have been. Thanks, too, to John's able assistant Wickliffe Hallos, who rode herd on this book through the bewildering (and sometimes maddening) publishing process. Many thanks also to W. W. Norton for bringing this book out into the world with so much enthusiasm.

I am indebted to Salisa Rabinowitz who put her trust in me, generously providing me with so many important details and bravely dredging up the still-painful memories of her husband's death; to

Jane Alexander, who amiably offered her insights into a charismatic and complicated man who was also her dear friend; and to the magnanimous George Schaller who, while recovering from the death of his wife, patiently answered my many questions. And thanks to Miriam Horn, writer and filmmaker, who made my conversations with them possible. A very special thanks to Tom Kaplan, who joyously shared stories of his friendship with Alan and Panthera's early days.

I also need to thank Burns Ellison, fellow jaguar lover, wanderer, novelist, editor, and sidekick who was with me in spirit on so many of my adventures. Burns accompanied me on my first trip to Costa Rica all those years ago, when the seeds of this book were first planted as we looked out over the expanse of Corcovado National Park from our La Tarde porch. Though he couldn't travel the corridor with me, when the time came, he set aside his own work and devoted his time and skills to reading and editing versions of this book. For his companionship, thanks to my traveling pal and lifelong birder Ezra Garfield. Thanks, too, to friend Mike Boston, Irish raconteur and the Osa Peninsula's legendary guide, whose life was cut way too short.

This story has taken me many years to tell. My progress was interrupted by Covid, which made traveling difficult. And midway through, Howard Quigley, a champion of the book, passed away. Howard was loved and admired by all, and the jaguar community was devastated by his death.

At some point every book relies on serendipity, and I had the good fortune of being visited by it twice. After Howard's death, I struggled to regain my footing. That's when John Polisar, who spent much of his career at the WCS working at various points along the Jaguar Corridor, emerged as steadfast advocate of the book. A fount of knowledge with a razor-sharp mind and a wry wit, he guided me and gave freely and generously of his time, recommending essential papers and introducing me to biologists across the jaguar range (and to the blues rock performances he loved). This book may not be exactly the one he would have written or wanted me to write, but I hope it comes close.

My second experience with serendipity happened when my friend of three-plus decades, the estimable author David Gessner, agreed to join me in the desert Southwest on my backpacking adventure. Employing his cartoonist's skills, he sketched out my narrative for me, essentially leading a frustrated and stymied author to the crux of the story. Along the way, David and I conjured memories of our Colorado days and bonded once again over our mutual love of wild country.

The project has been a difficult and memorable one. Part of its beauty is having met so many people along the corridor that I've come to regard as friends. Chief among them is Joares May, who not only responded to my countless emails with patience and wisdom but along the way, became a pal and a passionate Green Bay Packer fan. I owe a tremendous amount of gratitude to him and to so many others whose thoughts inform this book. Whatever inaccuracies remain are mine alone.

To those whose contributions and kindnesses helped me immeasurably, a huge thank you: Carlos Manuel Rodríguez-Echandi, Roberto Salom-Pérez, Daniela Araya Gamboa, Daniel Corrales-Gutiérrez, Stephanie Mory Villaseñor, Erik Olson, Alejandro Azofeifa, Dani Herrera Badilla, Jason Cespedes, Stephanny Arroyo Arce, Guido Saborío, Eduardo Carrillo, Juan-Carlos Cruz, Steve Worley, John Podson and Rico at Cabinas Jiménez, Joost Wilms at Dantica Cloud Forest Lodge, Eduardo at La Tarde, Johannes Stahl, Juan Carlos Vasquez Murillo, Melissa Arias, Nick McPhee, Saul Arias Cossio, José Luis Beltrán, Elizabeth Unger, Marcos Uzquiano, Eduardo Franco Berton, Duston Larsen, Anai Holzmann, Damián Rumiz, Mariano Arrien-Gomez, Patrick O'Connell, Angelo Ambrosini, Karen Wood, Jorge Salomão, Raíssa Sepulvida, Fernando Tortato, Abbie Martin, Andrea Pisaro, Valeria Boron, Rafael Hoogesteijn, Allison Devlin, Wai-Ming Wong, Petterson Silva, Caroline Leuchtenberger, Silvia and Zedequias, and all the wonderful staff at Jofre Velho Conservation Ranch, Eric Sanderson, Tony Povilitis, Shawn Larson, Kathy Zeller, Robert Wallace, Esteban Payán, Mathias Tobler, Carlos Durigan, Emiliano Ramalho, Sebastián Di Martino, Diana Friedeberg, Alejandro Jesus de la Cruz, Michael Steinberg, Renata Leite Pitman, Nathan Utrup at Yale University's Peabody Museum of Natural History Archives, Greg Manocherian, Andy Sabin, Randy Serraglio,

Russ McSpadden, Mike Stark, Chris Bugbee, Sergio Avila, Michael Robinson, Jim Heffelfinger, Clay Crowder, Melanie Culver, Jack and Anna Mary Childs, Warner Glenn and Kelly Kimbro, Roberto Wolf, Diana Hadley, Gerardo Ceballos, John Koprowski, Valer Clark, Valerie Gordon, Alejandro Ganesh Marín, Andrea Crosta, Brandon Thompson, Taylor Turner, Sebastian Kennerknecht, and April Kelly.

I'd like also to thank my father-in-law, Daggett Harvey, for his unfailing support of my travels and my books, even as he counsels me to "write faster."

And last of all to my wife Elizabeth, poet, fellow adventurer, and best friend. For her I reserve my most heartfelt thanks.

Notes

PREFACE

1. Les Line, "Scientist at Work / Alan Rabinowitz; Indiana Jones Meets His Match in Burma Rabinowitz," *New York Times*, August 3, 1999.
2. Bryan Walsh, "The Indiana Jones of Wildlife Protection," *Time*, January 10, 2008.
3. Alan Rabinowitz, "We Are All Wildlife," *On Being with Krista Tippett*, July 22, 2010.
4. Quoted in Adam Popescu, "The Last Days of the Big Cat Survivor," *Men's Journal*, October 17, 2018.
5. Alan Rabinowitz, *An Indomitable Beast: The Remarkable Journey of the Jaguar* (Island Press, 2014), 7.
6. J. Antonio de la Torre et al., "The Jaguar's Spots Are Darker than They Appear: Assessing the Global Conservation Status of the Jaguar (*Panthera onca*)," *Oryx* 52, no. 2 (2018): 300–15.
7. Rachel Carson, *Silent Spring* (Houghton Mifflin, 1962), chap. 2.

INTRODUCTION: THE ROCK STAR

1. *The Arizona Republic, Arizona Daily Star, Tucson Sentinel*, and a host of other newspapers and magazines, from Miami to Washington, D.C., to Los Angeles, reported on El Jefe's status.
2. Tony Davis, "Jaguar Roared, Clawed at Hunting Dogs in Arizona Encounter," *Arizona Daily Star*, November 22, 2011.
3. Will Rizzo, "Return of the Jaguar?" *Smithsonian Magazine*, December 2005.

CHAPTER 1: THE JAGUAR CRAZE

1. Nicholas J. Saunders, *People of the Jaguar: The Living Spirit of Ancient America* (Souvenir Press, 1991), is devoted to those who worship the big cat. Richard Mahler, *The Jaguar's Shadow: Searching for a Mythic Cat* (Yale University Press, 2009) also devotes considerable space to examining the jaguar's mythic significance.

2. A. Starker Leopold, *Wildlife of Mexico: The Game Birds and Mammals* (University of California Press, 1959).
3. Dan Flores, *Coyote America: A Natural and Supernatural History* (Basic Books, 2016).
4. Vividly described in Dan Flores, *American Serengeti: The Last Big Animals of the Great Plains* (University Press of Kansas, 2016).
5. John James Audubon, *Audubon and his Journals*, ed. Maria R. Audubon (Charles Scribner's Sons, 1897), vol. 2, entry for August 11, 1843.
6. Flores, *American Serengeti*, 6.
7. David E. Brown and Carlos López-González, *Borderland Jaguars: Tigres de la Frontera* (University of Utah Press, 2001), 93.
8. Flores, *American Serengeti*, 56–59.
9. David J. Schmidly, William E. Tydeman, and Alfred L. Gardner, eds., *United States Biological Survey: A Compendium of Its History, Personalities, Impacts, and Conflicts* (Museum of Texas Tech University, 2016).
10. Ken Ross, *Pioneering Conservation in Alaska* (University Press of Colorado, 2017), chap. 13.
11. Flores, *Coyote America*, 123.
12. Rebecca Mead, "Should Leopards Be Paid for Their Spots?" *New Yorker*, March 21, 2022.
13. Rabinowitz, *Indomitable Beast*, 78–79; and Barbara Pascarell Brown, "Pretty in Pink: Jacqueline Kennedy and the Politics of Fashion," master's thesis, University at Albany, State University of New York, 2012.
14. Esteban Payán and Luis Trujillo, "The Tigrilladas in Colombia," *Cat News* 44 (2006).
15. Andre P. Antunes, "Historical Commercial Hunting of Mammals in Amazonia," in *Amazonian Mammals*, ed. W. R. Spironello et al. (Springer, 2023).
16. George Schaller, South America Field Notebooks and Journals, George Beals Schaller Collection, Yale Peabody Museum; George B. Schaller, *A Naturalist and Other Beasts: Tales from a Life in the Field* (Sierra Club Books, 2007), 72–73.
17. Paula Mackay, "Cockscomb: It's About Cats," *Wildlife Conservation*, November–December 2004, 30–34.

CHAPTER 2: BOLIVIAN COCAINE

1. Brown and López-González, *Borderland Jaguars*, 139.
2. Neil D'Cruze et al., "Characterizing Trade at the Largest Wildlife Market of Amazonian Peru," *Global Ecology and Conservation* 28 (2021): e01631.
3. Roberto Navia, "Fang trafficking to China is putting Bolivia's jaguars in jeopardy," *Mongabay*, January 26, 2018.
4. Eduardo Franco Berton, interview by the author; and Eduardo Franco Berton, "A Journey into a Black Market for Jaguar Body Parts in Latin America," *Earth Journalism Network*, August 31, 2018.
5. Sharon Wilcox, "Encountering El Tigre: Jaguars, Knowledge, and Discourse in the Western World, 1492–1945," Ph.D. diss., University of Texas at Austin, May 2014, 202–3.
6. Rabinowitz, *Indomitable Beast*, 78.
7. Rabinowitz, *Indomitable Beast*, 80.

8. Payán and Trujillo, "Tigrilladas in Colombia."
9. John Polisar, Charlotte Davies, Mariana da Silva, et al., "A Global Perspective on Trade in Jaguar Parts from South America," *Cat News*, Special Issue no. 16 (Winter 2023): 74–83.

CHAPTER 3: THE DREAM

1. Alan Rabinowitz, *Jaguar: One Man's Struggle to Establish the World's First Jaguar Preserve* (Island Press, 2000), 5 and 15.
2. Wilcox, "Encountering El Tigre," 189.
3. Wilcox, "Encountering El Tigre," 192.
4. Wilcox, "Encountering El Tigre," 192.
5. Schaller, *Naturalist*, 75; and Schaller, South America Field Notebooks and Journals.
6. Schaller, *Naturalist*, 74–75; and José Luiz Andrade Franco, José Augusto Leitão Drummond, and Fernanda Pereira de Mesquita Nora, "History of Science and Conservation of the Jaguar (*Panthera onca*) in Brazil," *Historia Ambiental Latino-americana y Caribeña* 8, no. 2 (2018): 50
7. Rabinowitz, *Jaguar*, 4–12, 218.
8. Rabinowitz, *Indomitable Beast*, 96.
9. Rabinowitz, *Jaguar*, 27.
10. Archie Carr, *The Windward Road: Adventures of a Naturalist on Remote Caribbean Shores* (Robert Hale, 1957), 27–28.
11. Rabinowitz, *Jaguar*, 49–50.
12. Alan Rabinowitz, *Beyond the Last Village: A Journey of Discovery in Asia's Forbidden Wilderness* (Island Press, 2001), 192.
13. Rabinowitz, *Beyond the Last Village*, 192; Schaller, *Naturalist*, 23.
14. Rabinowitz, *Jaguar*, 53–54.

CHAPTER 4: THE GOD OF DEATH

1. Rabinowitz, *Jaguar*, 63.
2. Rabinowitz, *Jaguar*, 67.
3. Rabinowitz, *Jaguar*, 69, 63.
4. Rabinowitz, *Indomitable Beast*, 296; Rabinowitz, *Chasing the Dragon's Tail: The Struggle to Save Thailand's Wild Cats* (Doubleday, 1991), 110.
5. Rabinowitz, *Jaguar*, 71.
6. Schaller, *Naturalist*, 20.
7. Rabinowitz, *Jaguar*, 82, 296.
8. Rabinowitz, *Jaguar*, 29.
9. Rabinowitz, *Indomitable Beast*, 1–2.
10. Rabinowitz, *Indomitable Beast*, 2–3.
11. In his dedication to *Indomitable Beast*, Rabinowitz thanks Salisa and his children, Alexander and Alana, for tolerating his "years of work and travel" so that he could "help be the voice for the big cats."
12. Schaller, *Naturalist*, 20.

CHAPTER 5: THE SURVIVOR

1. Mark Hallett and John M. Harris, *On the Prowl: In Search of Big Cat Origins* (Columbia University Press, 2020).
2. Charles Repenning, "Late Pliocene–Early Pleistocene Ecologic Changes in the Arctic Ocean Borderland," *U.S. Geological Survey Bulletin*, no. 2036 (U.S. Government Printing Office, 1994); and Dan O'Neill, *The Last Giant of Beringia: The Mystery of the Bering Land Bridge* (Basic Books, 2004), 11.
3. O'Neill, *Last Giant of Beringia*, 121–22.
4. Steven Mithen, *After the Ice: A Global Human History, 20,000–5000 BC* (Harvard University Press, 2006).
5. David Quammen, *Monster of God: The Man-Eating Predator in the Jungles of History and the Mind* (W. W. Norton, 2003), 6.
6. Craig Childs, *Atlas of a Lost World: Travels in Ice Age America* (Pantheon, 2018), 202.
7. Paul S. Martin, "Africa and Pleistocene Overkill," *Nature* 212, no. 5060 (1966): 339–42.
8. Rabinowitz, *Indomitable Beast*, 178, 21.

CHAPTER 6: THE HUNTER

1. Rabinowitz, *Indomitable Beast*, 168–69.
2. Holly Spanner, "Top 10: Which Animals Have the Strongest Bite?" *BBC Science Focus*, April 29, 2023.
3. Schaller, *Naturalist*, 67.
4. Hallett and Harris, *On the Prowl*, 19–46.
5. Tim Elmo Feiten, "Jakob von Uexküll's Concept of Umwelt," *Philosopher* 110, no. 1 (Winter 2022).
6. Schaller, *Naturalist*, 67
7. Natalie Angier, "At Last, Ready for Its Close-Up," *New York Times*, June 17, 2003.

CHAPTER 7: PEOPLE OF THE JAGUAR

1. Saunders, *People of the Jaguar*, 129.
2. Saunders, *People of the Jaguar*, 129.
3. Saunders, *People of the Jaguar*, 42.
4. David Freidel, Linda Schele, and Joy Parker, *Maya Cosmos: Three Thousand Years on the Shaman's Path* (William Morrow, 1993), 137.
5. Saunders, *People of the Jaguar*, 74.
6. Brown and López-González, *Borderland Jaguars*, 68.
7. Richard Perry, *The World of the Jaguar* (David & Charles, 1970), 102.
8. Rabinowitz, *Indomitable Beast*, 39.
9. Quoted in Saunders, *People of the Jaguar*, 70.
10. Though I consulted numerous other sources, Saunders, Mahler, and Rabinowitz provided many of the details for what follows.
11. Sean S. Sell, "The Chiapas Jaguar as Symbol of Maya *Resintencia*—Resistance and Intention," *Latin Americanist* 65, no. 1 (2021): 105–22.
12. Saunders, *People of the Jaguar*, 149.

CHAPTER 8: THE EDGE

1. Howard Quigley, interview by the author; and Schaller, Brazil Journal, 1980, August 4, 20, in Schaller, South America Field Notebooks.
2. Rabinowitz, *Indomitable Beast*, 91.
3. Rabinowitz, *Indomitable Beast*, 97.
4. Rabinowitz, *Indomitable Beast*, 95.
5. Rabinowitz, *Indomitable Beast*, 97.
6. Jane Alexander, interview by the author.
7. Rabinowitz, *Beyond the Last Village*, 123.
8. Rabinowitz, *Jaguar*, 134.
9. Rabinowitz, *Jaguar*, 292.
10. Rabinowitz, *Jaguar*, 280–81.
11. Rabinowitz, *Jaguar*, 279.
12. Rabinowitz, *Jaguar*, 177.
13. Rabinowitz, *Jaguar*, 211.
14. Rabinowitz, *Jaguar*, 212.
15. Tom Miller, "Jaguars and Jungle Obsessions," *Washington Post*, December 12, 1986.
16. Rabinowitz, *Jaguar*, 214.

CHAPTER 9: HE WHO RIDES A TIGER CANNOT DISMOUNT

1. Rabinowitz, *Jaguar*, 318.
2. Rabinowitz, *Jaguar*, 318.
3. Jane Alexander, interview by the author.
4. Rabinowitz, *Jaguar*, 314–15.
5. Rabinowitz, *Jaguar*, 218, 277.
6. Rabinowitz, *Jaguar*, 216.
7. Rabinowitz, *Jaguar*, 230.
8. Rabinowitz, *Jaguar*, 319.
9. Rabinowitz, *Jaguar*, 320–21.
10. Alan Rabinowitz, *Life in the Valley of Death: The Fight to Save Tigers in a Land of Guns, Gold, and Greed* (Island Press, 2008), 67.
11. Rabinowitz, *Life in the Valley of Death*, 67.
12. Rabinowitz, *Chasing the Dragon's Tail*, 3.
13. Rabinowitz, *Chasing the Dragon's Tail*, 12.
14. Rabinowitz, *Chasing the Dragon's Tail*, 213.
15. Rabinowitz, *Chasing the Dragon's Tail*, 15.
16. Rabinowitz, *Chasing the Dragon's Tail*, 23.

CHAPTER 10: EL TIGRE

1. Mahler, *Jaguar's Shadow*, 343–35.
2. Angier, "At Last, Ready for Its Close-Up."
3. Angier, "At Last, Ready for Its Close-Up."

CHAPTER 11: THE MAGIC

1. Edward Abbey, *The Journey Home: Some Words in Defense of the American West* (E. P. Dutton, 1977), 38.
2. John McPhee, "The Encircled River—1," *New Yorker*, April 24, 1977; McPhee, *Coming into the Country* (Farrar, Straus & Giroux, 1991), 62.
3. Rabinowitz, *Indomitable Beast*, 119.
4. Aldo Leopold, "A Biotic View of Land," *Journal of Forestry* 37, no. 9 (September 1, 1939): 727–30.
5. Aldo Leopold, *A Sand County Almanac: And Sketches Here and There* (Oxford University Press, 1949), 144.
6. Rabinowitz, *Indomitable Beast*, 71.

CHAPTER 12: THE MOTHER LODE

1. Erik Olson, J. B. Franke, et al., *Wildlife Monitoring Report for Corcovado National Park, Costa Rica, 2019* (Northland College, 2020); Erik Olson, G. Saborío, and J. C. Salazar, "Age of the Jaguar: A Unique Approach to Evaluating the Lifespan of a Rare Carnivore," *Cat News* 70 (2019): 36–38.

CHAPTER 13: A DREAM UNDONE?

1. Jane Alexander, *Wild Things, Wild Places: Adventurous Tales of Wildlife and Conservation on Planet Earth* (Knopf, 2016), 21.
2. Rabinowitz, *Indomitable Beast*, 116; Rabinowitz, *Chasing the Dragon's Tail*, 150.
3. Alexander, *Wild Things*, 21.
4. Rabinowitz, *Chasing the Dragon's Tail*, 87.
5. Alexander, *Wild Things, Wild Places*, 22.
6. Miller, "Jaguars and Jungle Obsessions."
7. Rabinowitz, *Jaguar*, xiv.
8. Alexander, *Wild Things, Wild Places*, 31.
9. Rabinowitz, *Beyond the Last Village*, 4.
10. Rabinowitz, *Life in the Valley of Death*, 92.
11. Rabinowitz, *Chasing the Dragon's Tail*, 90.
12. Rudyard Kipling, *From Sea to Sea and Other Sketches, Letter of Travel* (Doubleday, 1899); Rabinowitz, *Beyond the Last Village*, 43.
13. Rabinowitz, *Jaguar*, 360.
14. Rabinowitz, *Beyond the Last Village*, 188.
15. Line, "Scientist at Work."
16. Line, "Scientist at Work."
17. Rabinowitz, *Life in the Valley of Death*, 66.
18. Rabinowitz, *Life in the Valley of Death*, 82.

CHAPTER 14: A BOLD CONSERVATION MODEL

1. Rabinowitz, *Indomitable Beast*, 126–29.

2. Shawn E. Larson. "Taxonomic Re-Evaluation of the Jaguar," *Zoobiology* 16, no. 2 (1997): 107–20; Eizirik et al., "Phylogeography."
3. Rabinowitz, *Indomitable Beast*, 129.
4. Rabinowitz, *Indomitable Beast*, 129.
5. Rabinowitz, *Indomitable Beast*, 128.
6. Rabinowitz, *Indomitable Beast*, 130.
7. Rabinowitz, *Indomitable Beast*, 129.

CHAPTER 15: GREEN GOLD

1. Rabinowitz, *Indomitable Beast*, 116.
2. Dan Collyns, "'Murderer of Nature': Evo Morales Blamed as Bolivia Battles Devastating Fires," *Guardian*, September 2, 2019.
3. Alfredo Romero-Muñoz, Martin Jansen, et al., "Fires Scorching Bolivia's Chiquitano Forest," *Science* 366, no. 6469 (November 29, 2019): 1082.
4. Pauline Verheij, "An Assessment of Wildlife Poaching and Trafficking in Bolivia and Suriname," International Union for Conservation of Nature, 2019.

CHAPTER 16: LAND OF THIRST

1. Rosa Cuéllar et al., "Kaaiyana, a Jaguar with Cubs in the Kaa-Iya del Gran Chaco National Park, Bolivia," *Cat News* 57 (2012): 4–6.
2. Yvette Sierra Praeli, "Ñembi Guasu: Huge New Conservation Area in Bolivia's Gran Chaco," *Mongabay*, May 23, 2019.
3. Peter Matthiessen, *The Snow Leopard* (Viking Press, 1978), 223.

CHAPTER 17: FIELD OF DREAMS

1. Rabinowitz, *Life in the Valley of Death*, 74.
2. Rabinowitz, *Life in the Valley of Death*, 84.
3. Rabinowitz, *Life in the Valley of Death*, 147.
4. Rabinowitz, *Indomitable Beast*, 6.
5. Rabinowitz, *Indomitable Beast*, 120.
6. Rabinowitz, *Indomitable Beast*, 122.
7. Rabinowitz, *Indomitable Beast*, 122.

CHAPTER 18: THE JAGUAR CORRIDOR

1. Eric W. Sanderson, Kent H. Redford, et al., "Planning to Save a Species: The Jaguar as a Model," *Conservation Biology* 16, no. 1 (February 2002): 58–72.
2. Carlos Manuel Rodríguez-Echandi, interview by the author.
3. Rabinowitz, *Indomitable Beast*, 128.
4. H. B. Quigley and P. G. Crawshaw Jr., "A Conservation Plan for the Jaguar *Panthera onca* in the Pantanal Region of Brazil," *Biological Conservation* 61 (1992): 149–57.
5. Rabinowitz, *Indomitable Beast*, 148.

6. Katherine A. Zeller, Alan Rabinowitz, et al., "The Jaguar Corridor Initiative: A Range-Wide Conservation Strategy." In *Molecular Population Genetics, Evolutionary Biology, and Biological Conservation of Neotropical Carnivores*, ed. Manuel Ruiz-Garcia and Joseph M. Shostell (Nova Science Publishers, 2013).
7. Schaller, *Naturalist*, 24.
8. Rabinowitz, *Indomitable Beast*, 194.
9. Rabinowitz, *Indomitable Beast*, 191.

CHAPTER 19: THE BILLIONAIRE WHO LOVED BIG CATS

1. Rabinowitz, *Indomitable Beast*, 21, 138.
2. Ashlea Ebeling, "Tom Kaplan: Billionaire King of Cats," *Forbes*, October 27, 2013.
3. Schaller, *Naturalist*, 24; Steve Mirsky, "The Feral Biologist: A Talk with George Schaller," *Scientific American*, May 28, 2008.

CHAPTER 20: THE PANTANAL

1. Vic Banks, *The Pantanal: Brazil's Forgotten Wilderness* (Sierra Club Books, 1991).
2. Karl M. Wantzen et al., "The End of an Entire Biome? World's Largest Wetland, the Pantanal, Is Menaced by the Hidrovia Project Which Is Uncertain to Sustainably Support Large-Scale Navigation," *Science of the Total Environment* 908 (2024): 167751.
3. Rabinowitz, *Indomitable Beast*, 91.
4. Alexander, *Wild Things, Wild Places*, 71.
5. Peter Alexander et al., "Why the US–China Trade War Spells Disaster for the Amazon," *Nature*, March 27, 2019.
6. Theodore Roosevelt. *Through the Brazilian Wilderness* (Charles Scribner's Sons, 1914), 26.
7. Roosevelt, *Through Brazilian Wilderness*, 84.
8. Schaller, South America Field Notebooks and Journals, 79–80.

CHAPTER 21: ONÇA

1. Quoted in Perry, *World of Jaguar*, 137–38.
2. Quammen, *Monster of God*, 3, 329.
3. Matt Walker, "Jaguar Mums Give Up Baby Secrets," *Earth News*, May 29, 2009.
4. Leopold, *Wildlife of Mexico*, 49.
5. Rabinowitz, *Indomitable Beast*, 182.

CHAPTER 22: THE PYROCENE

1. Stephen Eisenhammer, "'Day of Fire': Blazes Ignite Suspicion in Amazon Town," Reuters, September 11, 2019.
2. Brent McDonald, "Threats and Promises in Brazil's Lawless Amazon," *New York Times*, October 6, 2019.
3. Mathias W. Tobler, Rony Garcia Anleu, et al., "Do Responsibly Managed Logging Concessions Adequately Protect Jaguars and Other Large and Medium-Sized

Mammals? Two Case Studies from Guatemala and Peru," *Biological Conservation* 220 (2018): 245–53.
4. J. J. Figel et al., "Overlooked Jaguar Guardians: Indigenous Territories and Range-Wide Conservation of a Cultural Icon." *Ambio* 51, no. 12 (December 2022): 2532–43.
5. Juliano A. Bogoni et al., "Impending Anthropogenic Threats and Protected Area Prioritization for Jaguars in the Brazilian Amazon," *Communications Biology* 6, no. 132 (2023).
6. Terrence McCoy, "The Amazon, Undone: A Failure of Enforcement, Deforesters Are Plundering the Amazon. Brazil Is Letting Them Get Away with It." *Washington Post*, August 30, 2022.
7. Ernesto Londoño, "Jair Bolsonaro, on Day 1, Undermines Indigenous Brazilians' Rights," *New York Times*, January 2, 2019.
8. Jake Spring and Anthony Boadle, "Brazil Indigenous Defender, Sidelined Under Bolsonaro, Gave Life for 'Abandoned' Tribes," Reuters, June 19, 2022; Conectas Human Rights, "Bolsonaro Presents a Retrograde 'New Brazil' at the UN," Conectas Human Rights, September 24, 2019.
9. Letícia Casado and Ernesto Londoño, "Under Brazil's Far-Right Leader, Amazon Protections Slashed and Forests Fall," *New York Times*, July 28, 2019.
10. Jenny Gonzales, "Brazil Bows to Pressure from Business, Decrees 120-Day Amazon Fire Ban," *Mongabay*, July 8, 2020.
11. Catrin Einhorn et al., "The World's Largest Tropical Wetland Has Become an Inferno," *New York Times*, October 13, 2020.
12. Stephen J. Pyne, *The Pyrocene: How We Created an Age of Fire, and What Happens Next*. (University of California Press, 2021).
13. Ana Ionova, "The World's Largest Wetland Is Burning, and Rare Animals Are Dying," *New York Times*, August 27, 2024.

CHAPTER 24: RETURN TO THE PANTANAL

1. Schaller, *Naturalist*, 92.
2. Rabinowitz, *Life in the Valley of Death*, 37.

CHAPTER 26: THE MAGIC CATS

1. Childs and Childs, *Ambushed on the Jaguar Trail*, 16.
2. U.S. Fish and Wildlife Service, "Endangered and Threatened Wildlife and Plants: Final Rule to Extend Endangered Status for the Jaguar in the United States," *Federal Register* 62, no. 140 (July 22, 1997), 39148 (Rules and Regulations).
3. USFWS, "Endangered and Threatened Wildlife and Plants," 39147–57.
4. USFWS, "Endangered and Threatened Wildlife and Plants," 39147–57.
5. USFWS, "Endangered and Threatened Wildlife and Plants," 39147–57.
6. USFWS, "Endangered and Threatened Wildlife and Plants," 39147–57.
7. Tony Davis, "Macho B: Death of a Rare Arizona Jaguar," *Arizona Daily Star*, June 22, 2010.
8. Jaguar Conservation Team, Summary Notes, Baxter Civic Center, Lordsburg,

NM, July 30, 1997, https://azgfd-portal-wordpress-pantheon.s3.amazonaws.com/wp-content/uploads/archive/1997-July-30.pdf.
9. Alan R. Rabinowitz, "The Present Status of Jaguars (*Panthera onca*) in the Southwestern United States," *Southwestern Naturalist* 44, no. 1 (March 1999): 96–100.
10. Jaguar Conservation Team, Summary Notes, Animas High School, New Mexico, July 31, 2003, https://azgfd-portal-wordpress-pantheon.s3.amazonaws.com/wp-content/uploads/archive/2003-July-31.pdf.
11. Steller, "Jaguar Team Ceases Work."
12. Emil McCain and Jack Childs, "Evidence of Resident Jaguars (*Panthera onca*) in the Southwestern United States and the Implications for Conservation," *Journal of Mammalogy* 89 (2008): 1–10.
13. McCain and Childs, "Evidence."
14. Steller, "Jaguar Team Ceases Work."

CHAPTER 27: THE TRAGIC CAT

1. Janay Brun, *Cloak and Jaguar: Following a Cat from Desert to Courtroom*, 2nd ed. (published by the author, November 2, 2018), 103, 115.
2. Brun, *Cloak and Jaguar*, 119–25; Dennis Wagner, "The Cat, the Captors and the Cover-Up," *Arizona Republic*, December 9, 2012.
3. Davis, "Macho B: Death of a Rare Jaguar."
4. Tim Steller, "Arizona Leaves Investigation of Jaguar's Capture, Death to Feds," *Arizona Daily Star*, April 30, 2009.
5. Shaun Slifer, "Whistling for Macho B: an Interview w/ Janay Brun," part 1, *Just-Seeds*, May 8, 2012.
6. Wagner, "Cat, the Captors."
7. *Arizona Capitol Reports* Staff, "No Jaguars Left in U.S. After Death of Macho B," *Arizona Capitol Times*, March 6, 2009.
8. Tony Davis, "Did Jaguar Macho B Have to Die?" *Arizona Daily Star*, March 29, 2009.
9. Arthur H. Rotstein, "Judge Orders Jaguar Plan by Jan. 8," *Arizona Daily Star*, April 1, 2009.
10. Steller, "Jaguar Team Ceases Work."
11. Davis, "Macho B: Death of a Rare Jaguar."
12. Tony Davis and Tim Steller, "State's Capture of Jaguar Macho B Was Intentional, Federal Investigators Conclude," *Arizona Daily Star*, June 21, 2010.
13. Alan Rabinowitz, "Jaguars Don't Live Here Anymore," *New York Times*, January 24, 2010.
14. Rabinowitz, *Indomitable Beast*, 158–61.
15. Wagner, "Cat, the Captors."
16. Wagner, "Cat, the Captors."
17. Brun, *Cloak and Jaguar*, 153.

CHAPTER 28: LAND OF THE JAGUAR

1. Nina Strochlic, "An Unlikely Feud Between Beekeepers and Mennonites Simmers in Mexico," *National Geographic*, April 12, 2019.
2. Jack Phillips, "Controversial Maya Train Up and Running as Mexicans Cast Their Ballots," North American Congress on Latin America, May 31, 2024.

CHAPTER 29: AMLO'S IRON HORSE

1. Alejandra García Quintanilla et al., "Impact of the Tren Maya Megaproject on the Biocultural Heritage of the Mayan Area in Mexico's Best Conserved Tropical Forest," *International Journal of Environmental Sciences and Natural Resources* 31, no. 3 (2022): 556317.
2. Thelma Gómez Durán, "'What's Lacking Is Respect for Mayan Culture': Q&A with Pedro Uc Be on Mexico's Tren Maya," *Mongabay*, May 30, 2022.
3. Gerard Soler, "Tren Maya, the Mexican Megaproject Threatening the Ecosystems of the Yucatán Peninsula," *Equal Times*, March 18, 2022.

CHAPTER 30: A SPECIES ON THE BRINK

1. Melissa Arias, "The Illegal Trade in Jaguars (*Panthera onca*)." Convention on International Trade in Endangered Species of Wild Fauna and Flora, May 7, 2021.
2. Arias, *Illegal Trade in Jaguars*, 7.
3. Arias, *Illegal Trade in Jaguars*, 57.
4. John Polisar, Charlotte Davies, Thaís Morcatty, et al., "Multi-Lingual Multi-Platform Investigations of Online Trade in Jaguar Parts." *PLOS One* 18, no. 1 (2023): e0280039.
5. Melissa M. Arias Goetschel, "Illegal Jaguar Trade in Latin America: An Evidence-Based Approach to Support Conservation Actions," Ph.D. thesis, University of Oxford, 2021, 175.
6. Barbara Fraser, "China's Lust for Jaguar Fangs Imperils Big Cats," *Nature* 555, no. 7694 (March 1, 2018): 13–14.
7. Brook Larmer, "Chinese Contradictions on Animal Trafficking," *Irish Examiner*, December 7, 2018.
8. Environmental Investigation Agency, "24 Firms in China Exposed for Using Bones of Endangered Leopards in Traditional Medicines," Environmental Investigation Agency, April 3, 2020.
9. Thaís Q. Morcatty et al., "Illegal Trade in Wild Cats and Its Link to Chinese-Led Development in Central and South America," *Conservation Biology* 34, no. 6 (2020): 1525–35.
10. Arias, *Illegal Trade in Jaguars*.
11. Larmer, "Chinese Contradictions."

CHAPTER 31: A COMING CRISIS

1. Quoted in Tad Friend, "Earth League International Hunts the Hunters," *New Yorker*, May 15, 2023.

2. Alexander, *Wild Things, Wild Places*, 24.
3. Christopher Jasparro, "Wildlife Trafficking and Poaching: Contemporary Context and Dynamics for Security Cooperation and Military Assistance," *CIWAG Case Studies* 17 (2018).
4. Rachel Nuwer, "Where Jaguars Are Killed, New Common Factor Emerges: Chinese Investment," *New York Times*, June 11, 2020.
5. Elizabeth L. Bennett, "Wildlife Trafficking's New Front: Latin America," Wildlife Conservation Society, October 11, 2018.

CHAPTER 32: THE BEAUTIFUL CAT

1. Sabrina Kenoun, "Potential Jaguar Habitat at U.S.-Mexico Border Identified by UA Researchers," *Tucson Sentinel*, April 19, 2021.
2. Leopold, *Wildlife of Mexico*, 464.

CHAPTER 33: JAGUAR INTERRUPTED

1. Russ McSpadden, "Copper World Is Coming for Your Beloved Santa Rita Mountains," *Arizona Daily Star*, July 1, 2019.
2. Rob Peters, cartoon, *Arizona Daily Star*, August 21, 2022.
3. Robert Peters et al., and 2,556 scientist signatories from 43 countries, "Nature Divided, Scientists United: US–Mexico Border Wall Threatens Biodiversity and Binational Conservation," *BioScience* 68, no. 10 (October 2018): 740–43.

CHAPTER 34: BRINGING JAGUARS HOME

1. Eric W. Sanderson, Jon P. Beckmann, et al., "The Case for Reintroduction: The Jaguar (*Panthera onca*) in the United States as a Model," *Conservation Science and Practice* 3, no. 6 (June 2021).
2. Tony Povilitis, *Jaguar Habitat Campaign* (blog).
3. Sanderson, Beckmann, et al., "Case for Reintroduction."
4. Adrian Hedden, "New Mexico Excluded from Federal Jaguar Recovery Efforts. Habitat Remains in Arizona," *Carlsbad Current-Argus*, July 24, 2021.
5. Center for Biological Diversity, "Petition to Reintroduce Jaguar (*Panthera onca*) to New Mexico and Designate Additional Critical Habitat in Arizona and New Mexico," Center for Biological Diversity, December 12, 2022.
6. Jason Plautz, "The Controversial Plan to Bring Jaguars Back to the US," *Vox*, October 12, 2021.
7. Quoted in Erin Stone, "Arizona, New Mexico Could Support More Jaguars in a Wider Area, a New Study Finds," *Arizona Republic*, March 17, 2021.
8. Sam Matey, "Innovative Wildlife Conservation," *Weekly Anthropocene: A Dispatch from the Wild, Weird World of Humanity and Its Biosphere*, November 30, 2022.
9. Leopold, *Sand County Almanac*, 143.
10. Quoted in Douglas Main, "The U.S. Border Wall Is Tearing Through Wilderness, Right Under Our Noses," *National Geographic*, November 2, 2020.
11. Center for Biological Diversity, "Petition to Reintroduce Jaguar," 91.

EPILOGUE

1. Jason Miller Outdoors. "Jaguar: An Arizona Trail Cam Journey" (video), Jason Miller Outdoors, January 3, 2024, https://www.youtube.com/watch?v=V6tNTKwMUAc.
2. Sillars Jordan, "Will Jaguars Be Reintroduced in the U.S.?" *MeatEater*, August 12, 2021.
3. Tony Davis, "U.S. Says It Won't Reintroduce Jaguars," *Arizona Daily Star*, January 24, 2024.
4. Panthera, "Jaguars in the United States: Your Questions, Answered," Panthera, January 23, 2024.
5. Global Conservation Corps, *Voices of Nature* (podcast), episode 3: "Allison Devlin Takes Us into the Pantanal."
6. Rabinowitz, *Indomitable Beast*, 178.

Sources

Abbey, Edward. *The Journey Home: Some Words in Defense of the American West.* E.P. Dutton, 1977.
Abhat, Divya. "Fenced Out: Wildlife Impacts of the US–Mexico Border Fence." *Wildlife Professional* 5, no. 22 (2011): 22–27.
Abi-Habib, Maria. "How a Tourist Paradise Became a Drug-Trafficking Magnet." *New York Times,* September 15, 2024.
Alberts, Elizabeth Claire. "For the Pantanal's Jaguars, Fires Bring 'Death by a Thousand Needle Wounds.'" *Mongabay,* September 24, 2020.
Alexander, Jane. *Wild Things, Wild Places: Adventurous Tales of Wildlife and Conservation on Planet Earth.* Knopf, 2016.
Alexander, Peter, et al. "Why the US–China Trade War Spells Disaster for the Amazon." *Nature,* March 27, 2019.
Amos, Amy Matthews. "Jaguars Thrive in Lightly Logged Forests." *Scientific American,* June 19, 2018.
Andean Information Network. "Debunking Myths: The Eastern Lowlands of Santa Cruz: Part of an Integrated Bolivia." Andean Information Network, January 29, 2010.
Andersen, Ross. "The Search for America's Atlantis." *Atlantic,* September 7, 2021.
Anderson, Jon Lee. "After Bolsonaro: Can Lula Remake Brazil?" *New Yorker,* January 30, 2023.
Andreoni, Manuela, and Ernesto Londoño. "Bolsonaro's Sudden Pledge to Protect the Amazon Is Met with Skepticism." *New York Times,* April 21, 2021.
Angier, Natalie. "At Last, Ready for Its Close-Up." *New York Times,* June 17, 2003.
Antunes, Andre P. "Historical Commercial Hunting of Mammals in Amazonia." In *Amazonian Mammals,* ed. W. R. Spironello et al. Springer, 2023.
Antunes, Claudia. "Luciana Gatti: The Amazon Is No Longer Offsetting Human Destruction." *Sumaúma,* August 23, 2023.
Arias, Melissa. "The Illegal Trade in Jaguars (*Panthera onca*)." Convention on International Trade in Endangered Species of Wild Fauna and Flora, May 7, 2021.
Arias, Melissa, Peter Coals, et al. "Reflecting on the Role of Human-Felid Conflict and Local Use in Big Cat Trade." *Conservation Science and Practice* 6, no. 1 (2024): e13030.

Arias, Melissa, Amy Hinsley, and E. J. Milner-Gulland. "Characteristics of, and Uncertainties About, Illegal Jaguar Trade in Belize and Guatemala." *Biological Conservation* 250 (2020): 108765.

———. "Use of Evidence for Decision-Making by Conservation Practitioners in the Illegal Wildlife Trade." *People and Nature* 3 no. 5 (2021): 1110–26.

Arias, Melissa, Amy Hinsley, Paola Nogales-Ascarrunz, et al. "Prevalence and Characteristics of Illegal Jaguar Trade in North-Western Bolivia." *Conservation Science and Practice* 3, no. 7 (2021): e444.

Arias Goetschel, Melissa M. "Illegal Jaguar Trade in Latin America: An Evidence-Based Approach to Support Conservation Actions." PhD thesis, University of Oxford, 2021.

Arida, Anna L., and Daniel Wilkinson. "Letter on the Amazon and its Defenders to the Organisation for Economic Cooperation and Development (OECD)." Human Rights Watch, January 27, 2021.

Arizona Capitol Reports Staff. "No Jaguars Left in U.S. After Death of Macho B." *Arizona Capitol Times*, March 6, 2009.

Arizona Game and Fish Department. *Arizona Jaguar Conservation Team*. Arizona Memory Project, 1997.

———. *Jaguar Report*, July 30, 1997. Arizona Game and Fish Department Archives.

———. *Jaguar Report*, July 31, 2003. Arizona Game and Fish Department Archives.

———. *Jaguar Conservation Team Summary Notes for August Meeting*, 2004. Arizona Game and Fish Department Archives.

Arruda, Victor. "#PrayforAmazonia: How the World's Largest Rainforest Is Walking Towards Its Point of No Return." *Harvard Political Review*, February 2, 2022.

Audubon, John James. *Audubon and his Journals*, ed. Maria R. Audubon, 2 vols. (Charles Scribner's Sons, 1897).

Babb, Randall, et al. "Updates of Historic and Contemporary Records of Jaguars (*Panthera onca*) from Arizona." *Journal of the Arizona-Nevada Academy of Science* 49, no. 2 (2022): 65–91.

Banks, Vic. *The Pantanal: Brazil's Forgotten Wilderness*. Sierra Club Books, 1991.

Barbara, Vanessa. "Bolsonaro Said His 'Specialty Is Killing.' He's Been True to His Word." *New York Times*, March 31, 2022.

Barnett, Lindsay. "Arizona Jaguar's Death Probably Hastened by Capture, Zoo Veterinarian Says." *Los Angeles Times*, March 4, 2009.

Beatley, Meaghan, and Sam Edwards. "Is Mexico's Mayan Train a Boondoggle?" *Nation*, May 22, 2020.

Bennett, Elizabeth L. "Wildlife Trafficking's New Front: Latin America." Wildlife Conservation Society, October 11, 2018.

Berenguer, Erika, et al. "Tracking the Impacts of El Niño Drought and Fire in Human-Modified Amazonian Forests." *Proceedings of the National Academy of Sciences of the United States of America* 118, no. 30 (2021): e2019377118.

Berton, Eduardo Franco. "A Journey into a Black Market for Jaguar Body Parts in Latin America." *Earth Journalism Network*, August 31, 2018.

Betts, Richard. "Hothouse Earth: Here's What the Science Actually Does—and Doesn't—Say." *Conversation*, August 9, 2018.

Bjork-James, Carwil. *Mass Protest and State Repression in Bolivian Political Culture:*

Putting the Gas War and the 2019 Crisis in Perspective. Harvard Human Rights Program, May 2020.

Black, Riley. "How Jaguars Survived the Ice Age." *National Geographic*, January 27, 2016.

Bogoni, Juliano A., et al. "Impending Anthropogenic Threats and Protected Area Prioritization for Jaguars in the Brazilian Amazon." *Communications Biology* 6, no. 132 (2023).

Bolsonaro, Jair. Inauguration speech (video), January 1, 2019, https://www.youtube.com/watch?v=FoezQ9JPf_s.

Boron, Valeria, et al. "Jaguar Densities Across Human-Dominated Landscapes in Colombia: The Contribution of Unprotected Areas to Long Term Conservation." *PLOS One* 11, no. 5 (2016): e0153973.

Boulhosa, Ricardo L. P., and F. C. C. Azevedo. "Perceptions of Ranchers Towards Livestock Predation by Large Felids in the Brazilian Pantanal." *Wildlife Research* 41 (2014): 356–65.

Boulton, C. A., T. M. Lenton, and N. Boers. "Pronounced Loss of Amazon Rainforest Resilience since the Early 2000s." *Nature Climate Change* 12 (2022): 271–78.

Boydston, Erin E., and Carlos A. Lopez Gonzalez. "Sexual Differentiation in the Distribution Potential of Northern Jaguars (*Panthera onca*)." In *Connecting Mountain Islands and Desert Seas: Biodiversity and Management of the Madrean Archipelago II*, ed. Gerald J. Gottfried et al., 51–56. Proceedings RMRS-P-36. U.S. Department of Agriculture, Forest Service, Rocky Mountain Research Station, 2005.

Branford, Sue. "From Nothing to Nowhere—The Trans Amazonian Highway." *New Internationalist*, October 2, 1980.

Brinkhof, Tim. "Clovis Debunked: America's First Settlers Did Not Take the Ice-Free Corridor." *Big Think*, April 11, 2022.

Brown, David E., and Carlos López-González. *Borderland Jaguars: Tigres de la Frontera*. University of Utah Press, 2001.

Brown, Barbara Pascarell. "Pretty in Pink: Jacqueline Kennedy and the Politics of Fashion." M.A. thesis, University at Albany, State University of New York, 2012.

Brun, Janay. *Cloak and Jaguar: Following a Cat from Desert to Courtroom*, 2nd ed. Published by the author, November 2, 2018.

———. *Whistling for the Jaguar: The Un-Redacted Story of the Jaguar, Macho B's Snaring and Death* (blog), https://whistlingforthejaguar.wordpress.com.

Buntin, Simmons. "Borderland Dreams: Tracking El Tigre in Southern Arizona and Beyond." Terrain.org, n.d.

Burd, Andrew. "Investigative Report on Macho B: The Last Known Living Jaguar in Arizona." *Climbing*, March 2018.

Campbell, David. *A Land of Ghosts: The Braided Lives of People and the Forest in Far Western Amazonia*. Houghton Mifflin, 2005.

Carr, Archie. *The Windward Road: Adventures of a Naturalist on Remote Caribbean Shores*. Robert Hale, 1957.

Carrara, Aline Fabiana Angotti. "The Struggle for Indigenous Territory in the Brazilian Amazon." Ph.D. diss., University of Florida, 2020.

Carrillo, Eduardo, Grace Wong, and Alfredo D. Cuarón. "Monitoring Mammal

Populations in Costa Rican Protected Areas Under Different Hunting Restrictions." *Conservation Biology* 14, no. 6 (2000): 1580–91.
Carson, Rachel. *Silent Spring*. Houghton Mifflin, 1962.
Casado, Letícia, and Ernesto Londoño. "Under Brazil's Far-Right Leader, Amazon Protections Slashed and Forests Fall." *New York Times*, July 28, 2019.
Cavalcanti, Sandra, and Eric Gese. "Kill Rates and Predation Patterns of Jaguars (*Panthera onca*) in the Southern Pantanal, Brazil." *Journal of Mammalogy* 91, no. 3 (2010).
Ceballos, Gerardo, Heliud Zarza, et al. "Beyond Words: From Jaguar Population Trends to Conservation and Public Policy in Mexico." *PLOS One* 16, no. 10 (2021): e0255555.
Ceballos, Gerardo, C. Chávez, et al., eds. *Jaguar Conservation and Management in Mexico: Case Studies and Perspectives*. Alianza WWF/Telcel-Universidad Nacional Autónoma de México, Mexico, 2011.
Center for Biological Diversity. "In-Danger Designation Requested for Pantanal Wetlands in Brazil, Bolivia, Paraguay." Center for Biological Diversity, February 2, 2022.
———. "Petition to Reintroduce Jaguar (*Panthera onca*) to New Mexico and Designate Additional Critical Habitat in Arizona and New Mexico." Center for Biological Diversity, December 12, 2022.
Charity, Sandra, and Juliana Machado Ferreira. *Wildlife Trafficking in Brazil*. Traffic International, July 2020.
Chaves, Leandro. "Luciana Gatti: 'We Are Wrecking Our Rain-Making Factory.'" InfoAmazonia, July 29, 2022.
Childs, Craig. *Atlas of a Lost World: Travels in Ice Age America*. Pantheon, 2018.
Childs, Jack L., and Anna Mary Childs. *Ambushed on the Jaguar Trail: Hidden Cameras on the Mexican Border*. Rio Nuevo, 2008.
Clark, Jorie, Anders E. Carlson, Alberto V. Reyes, and Dylan H. Rood. "The Age of the Opening of the Ice-Free Corridor and Implications for the Peopling of the Americas." *Proceedings of the National Academy of Sciences* 119, no. 14 (March 21, 2022): e2118558119.
Clark, Tim, Murray Rutherford, and Denise Casey, eds. *Coexisting with Large Carnivores: Lessons from Greater Yellowstone*. Island Press, 2005.
Collyns, Dan. "'Murderer of Nature': Evo Morales Blamed as Bolivia Battles Devastating Fires." *Guardian*, September 2, 2019.
Conectas Human Rights. "Bolsonaro Presents a Retrograde New Brazil at the UN." Conectas Human Rights, September 24, 2019.
Cormier, Jonathan. "Bolsonaro's Inauguration Speech: A New Vision for Brazil." *Medium*, January 1, 2024.
Cowan, Carolyn. "Fire and Forest Loss Ignite Concern for Brazilian Amazon's Jaguars." *Mongabay*, October 12, 2021.
Craighead, Kimberly, and Milton Yacelga. "Indigenous Peoples' Displacement and Jaguar Survival in a Warming Planet." *Global Sustainability* 4 (2021): 1–28.
Cuadros, Alex. "Has the Amazon Reached Its 'Tipping Point'?" *New York Times*, January 4, 2023.
Cuéllar, Rosa, et al. "Kaaiyana, a Jaguar with Cubs in the Kaa-Iya del Gran Chaco National Park, Bolivia." *Cat News* 57 (2012): 4–6.

Dalton, Jane. "Brazil Will Let Hunters Shoot Endangered Jaguars, Parrots, and Monkeys in Rainforests Under New Law, Warn Conservation Experts." *Independent*, August 13, 2019.
Darwin, Charles. *On the Origin of Species*. John Murray, 1859.
Dasgupta, Shreya. "At Least 500 Jaguars Lost in Amazon Fires." *Mongabay*, September 25, 2019.
Davis, Tony. "Death Won't Stop Jaguar Captures." *Arizona Daily Star*, March 6, 2006.
———. "Did Jaguar Macho B Have to Die?" *Arizona Daily Star*, March 29, 2009.
———. "Enviros Sue in Jaguar's Death." *Arizona Daily Star*, September 25, 2009.
———. "Jaguar Conservation Team Ceases Work." *Arizona Daily Star*, October 18, 2009.
———. "Macho B: Death of a Rare Arizona Jaguar." *Arizona Daily Star*, June 22, 2010.
———. "Jaguar Roared, Clawed at Hunting Dogs in Arizona Encounter," *Arizona Daily Star*, November 22, 2011.
———."Point-Counterpoint on Jaguar Critical Habitat." *Arizona Daily Star*, February 8, 2013.
———. "U.S. Says It Won't Reintroduce Jaguars." *Arizona Daily Star*, January 24, 2024.
Davis, Tony, and Tim Steller. "AZ Wanted a Jaguar Collared Despite Two Deaths in Sonora." *Arizona Daily Star*, June 4, 2009.
———. "I Baited Jaguar Trap, Research Worker Says." *Arizona Daily Star*, April 2, 2009.
———. "State's Capture of Jaguar Macho B Was Intentional, Federal Investigators Conclude." *Arizona Daily Star*, June 21, 2010.
Dayalu, Archana, et al. "Constraining 2010–2020 Amazonian Carbon Flux Estimates with Satellite Solar-Induced Fluorescence (SIF)." EGUsphere [preprint], April 24, 2024.
de la Torre, J. Antonio, et al. "The Jaguar's Spots Are Darker than They Appear: Assessing the Global Conservation Status of the Jaguar *Panthera onca*." *Oryx* 52, no. 2 (2018): 300–15.
de Landa, Friar Diego. *Yucatan: Before and After the Conquest*, translated by William Gates. Dover, 1978.
Demuth, Bathsheba. *Floating Coast: An Environmental History of the Bering Strait*. W. W. Norton, 2019.
Devlin, Allison L., et al. "Drivers of Large Carnivore Density in Non-Hunted, Multi-Use Landscapes." *Conservation Science and Practice* 5, no. 1 (January 6, 2023): e12745.
Díaz Santos, Fabricio. "Conserving Biological and Cultural Diversity in Nicaragua." *Medium*, March 3, 2020.
Duarte, Herbert O. B., et al. "Big Cats Like Water: Occupancy Patterns of Jaguar in a Unique and Insular Brazilian Amazon Ecosystem." *Mammal Research* 68, no. 3 (2023): 263–71.
Dummett, Cassie, and Arthur Blundell. *Illicit Harvest, Complicit Goods: The State of Illegal Deforestation for Agriculture*. Forest Policy Trade and Finance Initiative, May 2021.

Dumphreys, Lucas, and Diane Jeantet. "Fires in Brazil Threaten Jaguars, Houses, and Plants in the World's Largest Tropical Wetlands." Associated Press, November 19, 2023.

Durán, Thelma Gómez. "'What's Lacking Is Respect for Mayan Culture': Q&A with Pedro Uc Be on Mexico's Tren Maya." *Mongabay*, May 30, 2022.

Dussault, Antoine Corriveau. "Functionalism Without Selectionism: Charles Elton's 'Functional' Niche and the Concept of Ecological Function." *Biological Theory* 17, no. 1 (2022): 52–67.

D'Cruze, Neil, et al. "Characterizing Trade at the Largest Wildlife Market of Amazonian Peru." *Global Ecology and Conservation* 28 (2021): e01631).

Earth First! Journal of Ecological Resistance. Beltane 2011.

Ebeling, Ashlea. "Tom Kaplan: Billionaire King of Cats." *Forbes*, October 8, 2013.

Einhorn, Catrin, et al. "The World's Largest Tropical Wetland Has Become an Inferno." *New York Times*, October 13, 2020.

Eisenhammer, Stephen. "'Day of Fire': Blazes Ignite Suspicion in Amazon Town." Reuters, September 11, 2019.

Eizirik, Eduardo, et al. "Phylogeography, Population History, and Conservation Genetics of Jaguars (*Panthera onca, Mammalia, Felidae*)." *Molecular Ecology* 10, no. 1 (2001): 65–79.

Ellerbeck, Alexandra. "Bolivia's Aggressive Agricultural Development Plans Threaten Forests." *Mongabay*, June 8, 2015.

Enever, Andrew. "Bolivia Struggles to Halt Animal Trade." *BBC News*, March 5, 2002.

Environmental Investigation Agency. "24 Firms in China Exposed for Using Bones of Endangered Leopards in Traditional Medicines," Environmental Investigation Agency, April 3, 2020.

Escobar, Herton. "There's No Doubt That Brazil's Fires Are Linked to Deforestation, Scientists Say." *Science*, August 26, 2019.

Espinosa, Santiago, et al. "The Jaguar: Hope for a Brighter Future in the Americas." In *Imperiled: The Encyclopedia of Conservation*, ed. Dominick A. DellaSala and Michael I. Goldstein, 113–120. Elsevier, 2022.

Fahn, James. "Can Logging and Conservation Coexist?" *Earth Journalism Network*, May 18, 2012.

Farkas, João. "Amazônia Ocupada." Photographs, Tulane University Libraries.

Fascione, Nina, Aimee Delach, and Martin E. Smith. *People and Predators: From Conflict to Coexistence*. Island Press, 2004.

Feiten, Tim Elmo. "Jakob von Uexküll's Concept of Umwelt." *Philosopher* 110, no. 1 (Winter 2022).

Ferrante, L., and P. M. Fearnside. "Brazil's New President and 'Ruralists' Threaten Amazonia's Environment, Traditional Peoples, and the Global Climate." *Environmental Conservation* 46, no. 4 (2019): 261–63.

Figel, J. J., et al. "Overlooked Jaguar Guardians: Indigenous Territories and Range-Wide Conservation of a Cultural Icon." *Ambio* 51, no. 12 (December 2022): 2532–43.

Fishman, Andrew. "Brazil's Indigenous Groups Mount Unprecedented Protest Against Destruction of the Amazon." *Intercept*, August 28, 2021.

Flores, Bernardo M., et al. "Critical Transitions in the Amazon Forest System." *Nature* 626 (2024): 555–64.

Flores, Dan. *American Serengeti: The Last Big Animals of the Great Plains.* University Press of Kansas, 2016.

———. *Coyote America: A Natural and Supernatural History.* Basic Books, 2016.

Foster, R. J., et al. "Jaguar (*Panthera onca*) Density and Tenure in a Critical Biological Corridor." *Journal of Mammalogy* 101, no. 6 (2020): 1622–37.

Fountain, Henry. "Amazon Is Less Able to Recover from Droughts and Logging, Study Finds." *New York Times*, March 7, 2022.

Fox, Kara. "Defending the Amazon Is a Dangerous Undertaking. Critics Say Bolsonaro Is Making It Worse." CNN, June 13, 2022.

Franco, José Luiz Andrade, José Augusto Leitão Drummond, and Fernanda Pereira de Mesquita Nora. "History of Science and Conservation of the Jaguar (*Panthera onca*) in Brazil." *Historia Ambiental Latinoamericana y Caribeña* 8, no. 2 (2018): 42–72.

Fraser, Barbara. "China's Lust for Jaguar Fangs Imperils Big Cats." *Nature* 555, no. 7694 (March 1, 2018): 13–14.

Freidel, David, Linda Schele, and Joy Parker. *Maya Cosmos: Three Thousand Years on the Shaman's Path.* William Morrow, 1993.

Friend, Tad. "Earth League International Hunts the Hunters." *New Yorker*, May 15, 2023.

Gatti, Luciana V., et al. "Amazonia as a Carbon Source Linked to Deforestation and Climate Change." *Nature* 595 (2021): 388–93.

Gibson, Graeme. *The Bedside Book of Beasts.* Doubleday, 2009.

Global Conservation Corps. *Voices of Nature* (podcast).

Gonzales, Jenny. "Brazil Bows to Pressure from Business, Decrees 120-Day Amazon Fire Ban." *Mongabay*, July 8, 2020.

Graham-Rowe, Duncan. "Conservation in Myanmar: Under the Gun." *Nature* 435, no. 7044 (June 16, 2005): 870–72.

Greenfield, Patrick. "How to Rewild a Country: The Story of Argentina." *Guardian*, June 24, 2022.

Grigione, M. M., et al. "Identifying Potential Conservation Areas for Felids in the USA and Mexico: Integrating Reliable Knowledge across an International Border." *Oryx* 43, no. 1 (2009): 78–86.

Guynup, Sharon. "Brazil's Pantanal Is at Risk of Collapse, Scientists Say." *Mongabay*, December 20, 2022.

Hallett, Mark, and John M. Harris. *On the Prowl: In Search of Big Cat Origins.* Columbia University Press, 2020.

Hatten, James R., Annalaura Averill-Murray, and William E. Van Pelt. *Characterizing and Mapping Potential Jaguar Habitat in Arizona.* Technical Report no. 203. Nongame and Endangered Wildlife Program, Arizona Game and Fish Department, January 2003.

Hatten, John R., A. Averill-Murray, and William E. Van Pelt. "A Spatial Model of Potential Jaguar Habitat in Arizona." *Journal of Wildlife Management* 69, no. 3 (2005): 1024–33.

Hayward, Matt, et al. "Prey Preferences of the Jaguar (*Panthera onca*) Reflect the Post-Pleistocene Demise of Large Prey." *Frontiers in Ecology and Evolution* 3 (2016): 1–19.

Hedden, Adrian. "New Mexico Excluded from Federal Jaguar Recovery Efforts. Habitat Remains in Arizona." *Carlsbad Current-Argus*, July 24, 2021.

Hernández, Anuar D., et al. "Food Habits of Jaguar and Puma in a Protected Area and Adjacent Fragmented Landscape of Northeastern Mexico." *Natural Areas Journal* 35, no. 2 (2015): 308–17.

Hidalgo-Mihart, Mircea G., F. M. Contreras-Moreno, et al. "Validation of the Calakmul–Laguna de Terminos Corridor for Jaguars (*Panthera onca*) in South-Eastern Mexico." *Oryx* 52, no. 2 (2018): 292–99.

Hidalgo-Mihart, Mircea G., Alejandro Jesús-de la Cruz, et al. "Jaguar Density in a Mosaic of Disturbed/Preserved Areas in Southeastern Mexico." *Mammalian Biology* 98 (2019): 173–78.

Hisayasu, Louise. "Mediated Memory and the Internet: Indigenous Protagonism in Brazil." Master's thesis, University of California at Los Angeles, 2019.

Humboldt, Alexander von, and Aimé Bonpland. *Personal Narrative of Travels to the Equinoctial Regions of the New Continent During the Years 1799–1804*, translated by H. M. Williams, 4 vols. Longman, Hurst, Rees, Orme, & Brown, 1807/1814–19.

International Rivers. "Brazil's Legislative Assembly Approves Law Prohibiting Hydroelectric Dams on the Cuiabá River." International Rivers, May 5, 2022.

Introvigne, Massimo. "Tradition, Family and Property (TFP), and the Heralds of the Gospel: The Religious Economy of Brazilian Conservative Catholicism." *Alternative Spirituality and Religion Review* 7, no. 2 (2016): 245–60.

Ionova, Ana. "The World's Largest Wetland Is Burning, and Rare Animals Are Dying." *New York Times*, August 27, 2024.

Ionova, Ana, and Manuela Andreoni. "A Severe Drought Pushes an Imperiled Amazon to the Brink." *New York Times*, October 17, 2023.

Jaguar Conservation Team. Summary Notes, Animas High School, NM, July 31, 2003, https://azgfd-portal-wordpress-pantheon.s3.amazonaws.com/wp-content/uploads/archive/2003-July-31.pdf.

———. Summary Notes, Baxter Civic Center, Lordsburg, NM, July 30, 1997, https://azgfd-portal-wordpress-pantheon.s3.amazonaws.com/wp-content/uploads/archive/1997-July-30.pdf.

Jason Miller Outdoors. "A New 'Jaguar' in Arizona?" (video), Jason Miller Outdoors, January 3, 2024, https://www.youtube.com/watch?v=V6tNTKwMUAc.

Jasparro, Christopher. "Wildlife Trafficking and Poaching: Contemporary Context and Dynamics for Security Cooperation and Military Assistance." *CIWAG Case Studies* 17 (2018).

Jedrzejewski, Wlodzimierz, R. Hoogesteijn, et al. "Collaborative Behavior and Coalitions in Male Jaguars (*Panthera onca*)—Evidence and Comparison with Other Cats." *Behavioral Ecology and Sociobiology* 76 (2022): 121.

Jedrzejewski, Wlodzimierz, Ronaldo Morato, Nuno Negrões, et al. "Estimating Species Distribution Changes Due to Human Impacts: The 2020's Status of the Jaguar in South America." *Cat News*, Special Issue no. 16 (Winter 2023): 44–55.

Jedrzejewski, Wlodzimierz, Ronaldo Morato, Robert B. Wallace, et al. "Landscape Connectivity Analysis and Proposition of the Main Corridor Network for the Jaguar in South America." *Cat News*, Special Issue no. 16 (Winter 2023): 56–61.

Jelinek, Arthur J. "Man's Role in the Extinction of Pleistocene Faunas." In *Pleistocene Extinctions: The Search for a Cause*, ed. P. S. Martin and H. E. Wright, Jr. New Haven, CT: Yale University Press, 1967.

Johnson, Julia. "WATCH: Experts Say Jaguar Sightings Near Border Could Signal Return to US." *Washington Examiner*, August 17, 2022.

Johnson, Terry B., and William E. Van Pelt. *Conservation Assessment and Strategy for the Jaguar in Arizona and New Mexico*. Wildlife Management Division, Arizona Game and Fish Department, March 24, 1997.

Jordahl, Laiken. "Jaguars Once Thrived in Arizona. It's Time to Reintroduce This Apex Predator." *Arizona Republic*, January 16, 2024.

Jordan, Sillars. "Will Jaguars Be Reintroduced in the U.S.?" *MeatEater*, August 12, 2021.

Kamnitzer, Ruth. "When Nature Gives Them a Chance to Collab, Jaguars Aren't So Solitary After All." *Mongabay*, February 7, 2023.

Kantek, Daniel Luis Zanella, et al. "Jaguars from the Brazilian Pantanal: Low Genetic Structure, Male-Biased Dispersal, and Implications for Long-Term Conservation." *Biological Conservation* 259 (2021): 109153.

Keenan, Rodney. "Climate Change and Tropical Forests." In *Achieving Sustainable Management of Tropical Forests*, ed. Jürgen Blaser and Patrick D. Hardcastle. Burleigh Dodds Science Publishing, 2020.

Kelly, M. J., and S. Silver. "The Suitability of the Jaguar (*Panthera onca*) for Reintroduction." In *Reintroduction of Top-Order Predators*, ed. M. W. Hayward and M. J. Somers, 187–205. Wiley-Blackwell, 2009.

Kenoun, Sabrina. "Potential Jaguar Habitat at U.S.-Mexico Border Identified by UA Researchers." *Tucson Sentinel*, April 19, 2021.

Kimbrough, Liz. "Where to Save the Jaguars? Researchers Identify Top Ten Areas in the Brazilian Amazon." *Mongabay*, March 10, 2023.

Kipling, Rudyard. *From Sea to Sea and Other Sketches: Letters of Travel*. Doubleday, 1899.

Kroner, Rachel E. Golden, et al. "The Uncertain Future of Protected Lands and Waters." *Science* 364 (2019): 881–86.

Kruuk, Hans. *Hunter and Hunted: Relationships Between Carnivores and People*. Cambridge University Press, 2002.

Langewiesche, William. "The War for the Rainforest." *New York Times*, March 16, 2022.

Langlois, Jill. "Marina Silva on Brazil's Fight to Turn the Tide on Deforestation." *Yale Environment 360*, April 18, 2024.

Larmer, Brook. "China's Mixed Messages on the Global Trade in Endangered-Animal Parts." *New York Times Magazine*, November 27, 2018.

———. "Chinese Contradictions on Animal Trafficking." *Irish Examiner*, December 7, 2018.

Larson, Shawn E. "Taxonomic Re-Evaluation of the Jaguar." *Zoobiology* 16, no. 2 (1997): 107–20.

Leopold, Aldo. "A Biotic View of Land." *Journal of Forestry* 37, no. 9 (September 1, 1939).

———. *A Sand County Almanac: And Sketches Here and There*. Oxford University Press, 1949.

Leopold, A. Starker. *Wildlife of Mexico: The Game Birds and Mammals*. University of California Press, 1959.

Levers, Christian, et al. "Agricultural Expansion and the Ecological Marginalization of Forest-Dependent People." *Proceedings of the National Academy of Sciences* 118, no. 40 (2021): e2100436118.

Levy, Sharon. *Once and Future Giants: What Ice Age Extinctions Tell Us About the Fate of Earth's Largest Animals*. Oxford University Press, 2011.

Line, Les. "Scientist at Work/Alan Rabinowitz; Indiana Jones Meets His Match in Burma." *New York Times*, August 3, 1999.

Lloret, Rocío. "The Protected Area That Isn't: Bolivia's Ñembi Guasu Beset by Fires, Farms, Roads." *Mongabay*, October 17, 2022.

Londoño, Ernesto. "Bolsonaro Sworn In as Brazil's President, Cementing Rightward Shift." *New York Times*, January 1, 2019.

———. "Jair Bolsonaro, on Day 1, Undermines Indigenous Brazilians' Rights." *New York Times*, January 2, 2019.

Londoño, Ernesto, and Letícia Casado. "As Bolsonaro Keeps Amazon Vows, Brazil's Indigenous Fear 'Ethnocide.'" *New York Times*, April 19, 2020.

Loomis, Brandon. "A Border Wall Could Drive the Jaguar Extinct in America." *Arizona Republic*, May 23, 2018.

Lopez, Barry. *Arctic Dreams: Imagination and Desire in a Northern Landscape*. Scribner, 1986.

———. *Of Wolves and Men*. Scribner, 1978.

Lorimer, Jamie. *Wildlife in the Anthropocene: Conservation after Nature*. University of Minnesota Press, 2015.

Lovejoy, Thomas E., and Carlos Nobre. "Amazon Forest-to-Savannah Tipping Point Could Be Far Closer than Thought." *Mongabay*, March 5, 2018.

———. "Amazon Tipping Point: Last Chance for Action." *Science Advances* 5, no. 12 (2019): eaba2949.

MacDonald, David W., et al. "Trading Animal Lives: Ten Tricky Issues on the Road to Protecting Commodified Wild Animals." *BioScience* 71, no. 8 (2021): 846–60.

"Macho Uno: Tracking One of the Oldest Wild Jaguars." *Northland College News*, May 13, 2020.

Mackay, Paula. "Cockscomb: It's About Cats." *Wildlife Conservation*, November–December 2004, 30–34.

Macmillan, Leslie. "Hurdles Remain for Jaguar Habitat." *New York Times*, January 23, 2013.

Mahler, Richard. *The Jaguar's Shadow: Searching for a Mythic Cat*. Yale University Press, 2009.

Main, Douglas. "The U.S. Border Wall Is Tearing Through Wilderness, Right Under Our Noses." *National Geographic*, November 2, 2020.

Malhi, Yadvinder, et al. "Tropical Forests in the Anthropocene." *Annual Review of Environment and Resources* 39 (2014): 125–59.

Marengo, Jose A., Ana P. Cunha, et al. "Extreme Drought in the Brazilian Pantanal in 2019–2020: Characterization, Causes, and Impacts." *Frontiers in Water* 3 (February 23, 2021): 639204.

Marengo, Jose A., G. S. Oliveira, and L. M. Alves. "Climate Change Scenarios in the Pantanal." In *Dynamics of the Pantanal Wetland in South America*, ed. Ivan Bergier and Mario Luis Assine, 227–38. Springer, 2015.

Martin, Abigail. *Jaguar Field Guide 2021: Jaguar Identification Project*. Independently published, 2022.

Martin, Paul S. "Africa and Pleistocene Overkill." *Nature* 212, no. 5060 (1966): 339–42.
Matey, Sam. "Innovative Wildlife Conservation." *Weekly Anthropocene: A Dispatch from the Wild, Weird World of Humanity and Its Biosphere*, November 30, 2022.
Matthiessen, Peter, *The Snow Leopard*. Viking Press, 1978.
McCain, Emil, and Jack Childs. "Evidence of Resident Jaguars (*Panthera onca*) in the Southwestern United States and the Implications for Conservation." *Journal of Mammalogy* 89 (2008): 1–10.
McCoy, Terrence. "The Amazon, Undone: A Failure of Enforcement, Deforesters Are Plundering the Amazon. Brazil Is Letting Them Get Away with It." *Washington Post*, August 30, 2022.
McCullough, I., et al. "Integrating High-Resolution Remote Sensing and Empirical Wildlife Detection Data for Climate-Resilient Corridors Across Tropical Elevational Gradients." *Biological Conservation* 298 (2024): 110763.
McDonald, Brent. "Threats and Promises in Brazil's Lawless Amazon." *New York Times*, October 6, 2019.
McKenna, Phil. "The Jaguar Whisperer Who Gave Them a Voice." *New Scientist*, October 1, 2014.
McPhee, John. *Coming Into the Country*. Farrar, Straus & Giroux, 1991.
———. "The Encircled River—1." *New Yorker*, April 24, 1977.
McSpadden, Russ. "Copper World Is Coming for Your Beloved Santa Rita Mountains." *Arizona Daily Star*, July 1, 2019.
Mead, Rebecca. "Should Leopards Be Paid for Their Spots?" *New Yorker*, March 21, 2022.
Meissner, Rene R., et al. "Habitat Destruction Threatens Jaguars in a Mixed Land-Use Region of Eastern Bolivia." *Oryx* 58, no. 1 (2024): 110–20.
Mendoza, Patricia A., et al. "Domestic Networks Contribute to the Diversity and Composition of Live Wildlife Trafficked in Urban Markets in Peru." *Global Ecology and Conservation* 37 (2022): e02161.
Menezes, Jorge F. S., et al. "Deforestation, Fires, and Lack of Governance Are Displacing Thousands of Jaguars in Brazilian Amazon." *Conservation Science and Practice* 3, no. 8 (August 2021).
Miller, Tom. "Jaguars and Jungle Obsessions." *Washington Post*, December 12, 1986.
Miller Llana, Sarah. "Democracy Around the World Is Down, but Not Out. Test Case: Brazil." *Christian Science Monitor*, April 9, 2020.
Mirsky, Steve. "The Feral Biologist: A Talk with George Schaller." *Scientific American*, May 28, 2008.
Mithen, Steven. *After the Ice: A Global Human History, 20,000–5000 BC*. Harvard University Press, 2006.
Morato, Ronaldo G., et al. "Biology and Ecology of the Jaguar." *Cat News*, Special Issue no. 16 (Winter 2023): 6–13.
Morcatty, Thaís Q., et al. "Illegal Trade in Wild Cats and Its Link to Chinese-led Development in Central and South America." *Conservation Biology* 34, no. 6 (December 2020): 1525–35.
Navarro-Serment, Carlos J., C. A. López-González, and J. P. Gallo-Reynoso. "Occurrence of Jaguars (*Panthera onca*) in Sinaloa, Mexico." *Southwestern Naturalist* 50, no. 1 (2005): 102–5.

Nield, David. "What Triggered the Collapse of the Ancient Maya? A New Study Reads Like a Warning." *Science Alert*, July 19, 2022.

Nijhuis, Michelle. *Beloved Beasts: Fighting for Life in an Age of Extinction*. W. W. Norton, 2021.

Nijman, Vincent, et al. "Global Online Trade in Primates for Pets." *Environmental Development* 48 (2023): 100925.

Nobbs-Thiessen, Benjamin. "Soybeans and Milk: Community and Commodification in a Bolivian Mennonite Colony." *Anabaptist Historians*, September 29, 2016.

Noss, A. J., et al. "Comparison of Density Estimation Methods for Mammal Populations with Camera Traps in the Kaa-Iya del Gran Chaco Landscape." *Animal Conservation* 15 (2012): 527–35.

Nuwer, Rachel. "Where Jaguars Are Killed, New Common Factor Emerges: Chinese Investment." *New York Times*, June 11, 2020.

Olson, Erik, J. B. Franke, et al. *Wildlife Monitoring Report for Corcovado National Park, Costa Rica, 2019*. Northland College, 2020.

Olson, Erik, G. Saborío, and J. C. Salazar. "Age of the Jaguar: A Unique Approach to Evaluating the Lifespan of a Rare Carnivore." *Cat News* 70 (2019): 36–38.

Olson, Erik, Y. Shen, et al. "Conservation Crisis? Status of Jaguars *Panthera onca* in Corcovado National Park, Costa Rica." *Oryx* (in press).

O'Neill, Dan. *The Last Giant of Beringia: The Mystery of the Bering Land Bridge*. Basic Books, 2004.

Pallares, Gloria. "Tren Maya: The Good, the Bad and the Ugly of Mexico's Megaproject." Global Landscapes Forum, May 22, 2024.

Panthera. "Jaguars in the United States: Your Questions, Answered." *Panthera*, January 23, 2024.

Paredes Tamayo, Iván. "Indigenous Groups Call for Government Intervention as Land Grabbers Invade Bolivian Protected Area." *Mongabay*, November 12, 2021.

Paudel, Kumar, et al. "Evaluating the Reliability of Media Reports for Gathering Information About Illegal Wildlife Trade Seizures." *PeerJ*, April 5, 2022.

Payán, Esteban, Valeria Boron, et al. "Legal Status, Utilisation, Management and Conservation of the Jaguar in South America." *Cat News*, Special Issue no. 16 (Winter 2023) : 62–73.

Payán, Esteban, and Luis Trujillo. "The Tigrilladas in Colombia." *Cat News* 44 (2006).

Pelt, Bill Van, et al. Jaguar Conservation Team (JAGCT) Summary Notes, Douglas City Hall, Douglas, AZ, January 31, 2002.

Perez, Richard. "Deforestation of the Brazilian Amazon Under Jair Bolsonaro's Reign: A Growing Ecological Disaster and How It May Be Reduced." *University of Miami Inter-American Law Review* 52, no. 2 (2021): 193–224.

Pérez Ortega, Rodrigo, and Inés Gutiérrez Jaber. "Controversial Train Heads for Maya Rainforest." *Science*, January 19, 2022.

Perry, Richard. *The World of the Jaguar*. David & Charles, 1970.

Peters, Robert, et al., and 2,556 scientist signatories from forty-three countries. "Nature Divided, Scientists United: US–Mexico Border Wall Threatens Biodiversity and Binational Conservation." *BioScience* 68, no. 10 (October 2018).

Petracca, Lisanne S., and L. Hernández-Santín. "Occupancy Estimation of Jaguar

(*Panthera onca*) to Assess the Value of East-Central Mexico as a Jaguar Corridor." *Oryx* 48, no. 1 (2014): 133–40.

Phillips, Jack. "Controversial Maya Train Up and Running as Mexicans Cast Their Ballots." North American Congress on Latin America, May 31, 2024.

Pietsch, Bryan. "'El Jefe,' Famed Arizona Jaguar, Feared Dead, Spotted in Mexico." *Washington Post*, August 10, 2022.

Plautz, Jason. "The Controversial Plan to Bring Jaguars Back to the US." *Vox*, October 12, 2021.

Podgorny, Irina. "From Jaguar Teeth to the Nail of the Great Beast: The Evolution of Animal Medicines." *Conversation*, December 19, 2018.

Polisar, John, Charlotte Davies, Thaís Morcatty, et al. "Multi-Lingual Multi-Platform Investigations of Online Trade in Jaguar Parts." *PLOS One* 18, no. 1 (2023): e0280039.

Polisar, John, Charlotte Davies, Mariana da Silva, et al. "A Global Perspective on Trade in Jaguar Parts from South America." *Cat News*, Special Issue no. 16 (Winter 2023): 74–83.

Polisar, John, Almira Hoogesteijn, et al. "The Rich Tradition of Jaguar Research and Conservation in Venezuela and Its Impact on Human-Jaguar Coexistence throughout the Species' Range." *Anartia*, June 2022, 79–95.

Polisar, John, Timothy O'Brien, et al. "Review of Jaguar Survey and Monitoring Techniques and Methodologies." 2014.

Polisar, John, Benoit Thoisy, et al. "Using Certified Timber Extraction to Benefit Jaguar and Ecosystem Conservation." *Ambio* 46 (2016): 1–10.

Popescu, Adam. "The Last Days of the Big Cat Survivor." *Men's Journal*, October 17, 2018.

Povilitis, Tony. *Jaguar Habitat Campaign* (blog).

Povilitis, Tony. "Jaguar Recovery Plan: Comments on the US Fish & Wildlife Service Draft Document, March 20, 2017."

Povilitis, Tony. "Recovering the Jaguar Panthera Onca in Peripheral Range: A Challenge to Conservation Policy." *Oryx* 49, no. 4 (2015): 626–31.

Povilitis, Tony, and C. Dustin Becker. "Advancing Large Carnivore Recovery in the Southwest." In *Southwestern Desert Resources*, ed. William L. Halvorson, Cecil. R. Schwalbe, and Charles. van Riper III, 263–73. University of Arizona Press, 2010.

Praeli, Yvette Sierra. "Ñembi Guasu: Huge New Conservation Area in Bolivia's Gran Chaco." *Mongabay*, May 23, 2019.

Project RADAM. "The Amazon Has Lost All Subjectivity," translated by Peter Schmidt. *Baffler*, April 25, 2023.

Proulx, Annie. *Fen, Bog and Swamp: A Short History of Peatland Destruction and Its Role in the Climate Crisis*. Scribner, 2022.

Pyne, Stephen J. *The Pyrocene: How We Created an Age of Fire, and What Happens Next*. University of California Press, 2021.

Quammen, David. *Monster of God: The Man-Eating Predator in the Jungles of History and the Mind*. W. W. Norton, 2003.

Quigley, Howard B., and David Bergman. "Jaguar Recovery Plan Final." U.S. Fish and Wildlife Service, 2018.

Quigley, Howard B., and Peter Crawshaw Jr. "Use of Ultralight Aircraft in Wildlife Radio Telemetry." *Wildlife Society Bulletin* 17, no. 3 (1989): 330–34.

———. "A Conservation Plan for the Jaguar (*Panthera onca*) in the Pantanal Region of Brazil." *Biological Conservation* 61, no. 3 (1992): 149–57.

Quigley, Howard B., Esteban Payán, et al. "Prologue: Why Care About Jaguars?" *Cat News*, Special Issue no. 16 (Winter 2023): 3–5.

Quintana, Itxaso, et al. "Severe Conservation Risks of Roads on Apex Predators." *Scientific Reports* 12, no. 1 (2022).

Quintanilla, Alejandra García, et al. "Impact of the Tren Maya Megaproject on the Biocultural Heritage of the Mayan Area in Mexico's Best Conserved Tropical Forest." *International Journal of Environmental Sciences and Natural Resources* 31, no. 3 (2022): 556317.

Rabelo, Rafael M., Susan Aragón, and Júlio César Bicca-Marques. "Prey Abundance Drives Habitat Occupancy by Jaguars in Amazonian Floodplain River Islands." *Acta Oecologica* 97 (2019): 28–33.

Rabinowitz, Alan. *Beyond the Last Village: A Journey of Discovery in Asia's Forbidden Wilderness*. Island Press, 2001.

———. *Chasing the Dragon's Tail: The Struggle to Save Thailand's Wild Cats*. Doubleday, 1991.

———. *An Indomitable Beast: The Remarkable Journey of the Jaguar*. Island Press, 2014.

———. *Jaguar: One Man's Struggle to Establish the World's First Jaguar Preserve*. Island Press, 2000.

———. "Jaguars Don't Live Here Anymore." *New York Times*, January 24, 2010.

———. *Life in the Valley of Death: The Fight to Save Tigers in a Land of Guns, Gold, and Greed*. Island Press, 2008.

———. "A New Strategy for Saving the World's Wild Big Cats." *Yale Environment 360*, February 17, 2010.

———. "The Present Status of Jaguars (*Panthera onca*) in the Southwestern United States." *Southwestern Naturalist* 44, no. 1 (March 1999): 96–100.

———. "We Are All Wildlife," *On Being with Krista Tippett* (podcast), July 22, 2010.

———. *Wildlife Field Research and Conservation Training Manual*. Wildlife Conservation Society, 1993.

Radwin, Maxwell. "Full Steam Ahead for Tren Maya Project as Lawsuits Hit Judicial Hurdles." *Mongabay*, January 28, 2022.

———. "Miners, Drug Traffickers and Loggers: Is Costa Rica's Corcovado National Park on the Verge of Collapse?" *Mongabay*, June 20, 2022.

Repenning, Charles. "Late Pliocene–Early Pleistocene Ecologic Changes in the Arctic Ocean Borderland." *U.S. Geological Survey Bulletin* no. 2036 (U.S. Government Printing Office, 1994).

Revkin, Andrew. "Brazil's Amazon Forest Is in the Crosshairs, as Defenders Step Up." *National Geographic*, December 21, 2018.

Rivera, Christian J. "Facing the 2013 Gold Rush: A Population Viability Analysis for the Endangered White-Lipped Peccary (*Tayassu pecari*) in Corcovado National Park, Costa Rica." *Natural Resources* 5, no. 16 (2014).

Rizzo, Will. "Return of the Jaguar?" *Smithsonian Magazine*, December 2005.

Robbins, Jim. "Wildlife Personalities Play a Role in Nature." *New York Times*, March 5, 2022.

Robinson, Michael J. "Investigative Report on Macho B—The Last Known Living Jaguar in Arizona." *Climbing Magazine*, January 29, 2010.

———. *Suitable Habitat for Jaguars in New Mexico: Report to the Arizona Game and Fish Department*. Center for Biological Diversity, 2006.

Rocha, Jan. "Drought Bites as Amazon's Flying Rivers Dry Up." *Guardian*, September 15, 2014.

Romero-Muñoz, Alfredo, Álvaro Fernández-Llamazares, et al. "A Pivotal Year for Bolivian Conservation Policy." *Nature Ecology and Evolution* 3 (2019): 866–69.

Romero-Muñoz, Alfredo, Martin Jansen, et al. "Fires Scorching Bolivia's Chiquitano Forest." *Science* 366, no. 6469 (November 29, 2019): 1082.

Romero-Muñoz, Alfredo, Ronaldo G. Morato, et al. "Beyond Fangs: Beef and Soybean Trade Drive Jaguar Extinction." *Frontiers in Ecology and the Environment* 18, no. 2 (March 2020): 67–68.

Romo, Vanessa. "Poaching Pressure Mounts on Jaguars, the Americas' Iconic Big Cat." *Mongabay*, September 14, 2020.

Roosevelt, Theodore. *Through the Brazilian Wilderness*. Charles Scribner's Sons, 1914.

Ross, Ken. *Pioneering Conservation in Alaska*. University Press of Colorado, 2017.

Rotstein, Arthur H. "Judge Orders Jaguar Plan by Jan. 8." *Arizona Daily Star*, April 1, 2009.

Rumiz, Damián I., Kathia Rivero-Guzmán, and Luis H. Acosta, "Jaguar Fangs and Other Animal Parts Confiscated by Bolivian Authorities and Examined at the Museo Noel Kempff Mercado." *Kempffiana* 20, no. 2 (2024): 18–34.

Sachs, Wolfgang. "The Age of Development: An Obituary." *New Internationalist*, February 27, 2020.

Sanderson, Eric W. "Jaguar Habitat Rediscovered in Arizona and New Mexico." *Oryx*, March 16, 2021.

———. "Let's Rebuild the U.S. Jaguar Population: Yes, Jaguars." *Scientific American*, June 8, 2021.

Sanderson, Eric W., Jon P. Beckmann, et al. "The Case for Reintroduction: The Jaguar (*Panthera onca*) in the United States as a Model." *Conservation Science and Practice* 3, no. 6 (June 2021).

Sanderson, Eric W., K. Fisher, et al. "A Systematic Review of Potential Habitat Suitability for the Jaguar (*Panthera onca*) in Central Arizona and New Mexico, USA." *Oryx* 56, no. 1 (2022): 116–127.

Sanderson, Eric W., Kent H. Redford, et al. "Planning to Save a Species: The Jaguar as a Model." *Conservation Biology* 16, no. 1 (February 2002): 58–72.

Sandy, Matt. "Murder of Brazil Official Marks New Low in War on Amazon Environmentalists." *Guardian*, October 24, 2016.

Saunders, Nicholas J. *People of the Jaguar: The Living Spirit of Ancient America*. Souvenir Press, 1991.

Schaller, George B. *A Naturalist and Other Beasts: Tales from a Life in the Field*. Sierra Club Books, 2007.

———. South America Field Notebooks and Journals. George Beals Schaller Collection, Yale Peabody Museum.

Schele, Linda, and David Freidez. *A Forest of Kings: The Untold Story of The Ancient Maya*. William Morrow, 1990.

Schmidly, David J., William E. Tydeman, and Alfred L. Gardner, eds. *United States Biological Survey: A Compendium of Its History, Personalities, Impacts, and Conflicts*. Museum of Texas Tech University, 2016.

Schneider, Victoria. "Across Latin America, Mennonites Seek Out Isolation at the Expense of Forests." *Mongabay*, December 15, 2021.

Schwitzer, Matthew. "'Not One Centimeter': President Bolsonaro's War on Indigenous Rights." *Columbia Undergraduate Law Review*, November 7, 2019.

Sell, Sean S. "The Chiapas Jaguar as Symbol of Maya *Resintencia*—Resistance and Intention." *Latin Americanist* 65, no. 1 (2021): 105–22.

Seymour, Kevin L. "Panthera onca." *Mammalian Species* 340 (October 26, 1989): 1–9.

Shepard, Paul. *The Others: How Animals Made Us Human*. Island Press, 1996.

———. *Traces of an Omnivore*. Island Press, 1996.

Sillars, Jordan. "Will Jaguars Be Reintroduced in the U.S.?" *MeatEater*, August 12, 2021.

Silva-Caballero, Adrián. "Land Use Change and Its Implications for Biodiversity and Jaguar Conservation." *Therya* 13, no. 3 (2022): 277–86.

Skidmore, Allison. "Exploring the Motivations Associated with the Poaching and Trafficking of Amur Tigers in the Russian Far East." *Deviant Behavior* 44, no. 3 (2022): 331–58.

Skirka, Hayley. "Battling Fires in Brazil's Wetlands: Scorched Paws and Firefighting Scientists." *National*, November 2, 2020.

Slifer, Shaun, "Whistling for Macho B: An Interview w/ Janay Brun," part 1, *JustSeeds*, May 8, 2012; part 2, *JustSeeds*, May 9, 2012.

Soler, Gerard. "Tren Maya, the Mexican Megaproject Threatening the Ecosystems of the Yucatán Peninsula." *Equal Times*, March 18, 2022.

Spanner, Holly. "Top 10: Which Animals Have the Strongest Bite?" *BBC Science Focus*, April 29, 2023.

Spring, Jake, and Anthony Boadle. "Brazil Indigenous Defender, Sidelined Under Bolsonaro, Gave Life For 'Abandoned' Tribes." Reuters, June 19, 2022.

Spurrier, Lauren. "Jaguar Conservation Is Key to Safeguarding South America's Pantanal Wetlands." Pew Charitable Trusts, July 2, 2024.

Steller, Tim. "Arizona Leaves Investigation of Jaguar's Capture, Death to Feds." *Arizona Daily Star*, April 30, 2009.

———. "Jaguar Conservation Team: The Extra Cuts." *Arizona Daily Star*, October 20, 2009.

———. "Jaguar Team Ceases Work amid Disputes, Big Cat's Death." *Arizona Daily Star*, October 18, 2009.

Steller, Tim, and Tony Davis. "State's Capture of Jaguar Macho B Was Intentional, Federal Investigators Conclude." *Arizona Daily Star*, June 21, 2010.

Stokstad, Erik. "New Research on Amazon Forest Fires." *Science*, November 29, 2019.

Stone, Erin. "Arizona, New Mexico Could Support More Jaguars in a Wider Area, A New Study Finds." *Arizona Republic*, March 17, 2021.

Strochlic, Nina. "An Unlikely Feud Between Beekeepers and Mennonites Simmers in Mexico." *National Geographic*, April 12, 2019.

Tamburini, Leonardo. "Indigenous Governance and Conservation of the Commons in Bolivia." International Work Group for Indigenous Affairs, October 11, 2023.

Thompson, Jeffrey J., et al. "Jaguar Status, Distribution, and Conservation in South-Eastern South America." *Cat News*, Special Issue no. 16 (Winter 2023): 35–43.

Tobler, Mathias W., Rony Garcia Anleu, et al. "Do Responsibly Managed Logging Concessions Adequately Protect Jaguars and Other Large and Medium-Sized Mammals? Two Case Studies from Guatemala and Peru." *Biological Conservation* 220 (2018): 245–53.

Tobler, Mathias W., Samia E. Carrillo-Percastegui, et al. "High Jaguar Densities and Large Population Sizes in the Core Habitat of the Southwestern Amazon." *Biological Conservation* 159 (March 2013): 375–81.

Tomas, Walfrido M., et al. "Sustainability Agenda for the Pantanal Wetland: Perspectives on a Collaborative Interface for Science, Policy, and Decision-Making." *Tropical Conservation Science* 12, no. 1 (2020): 1–11.

Torres, Cirenia A. "Assessing the Connectivity for the Jaguar (*Panthera onca*) in the United States-Mexico Border Ecoregions Using Species Distribution Modeling and Factorial Least Cost Path Analysis." M.S. thesis, University of Southern California, August 2021.

Torres, Natalia, et al. "Jaguar Distribution in Brazil: Past, Present and Future." *Cat News*, Special Issue no. 4 (2008): 4–8.

Tortato, Fernando R., Rafael Hoogesteijn, et al. "Reconciling Biome-Wide Conservation of an Apex Carnivore with Land-Use Economics in the Increasingly Threatened Pantanal Wetlands." *Scientific Reports* 11 (2021): 22808.

Tortato, Fernando R., and Thiago J. Izzo. "Advances and Barriers to the Development of Jaguar-Tourism in the Brazilian Pantanal." *Perspectives in Ecology and Conservation* 15, no. 1 (2017): 61–63.

Tortato, Fernando R., Thiago J. Izzo, Rafael Hoogesteijn, and Carlos A. Peres. "The Numbers of the Beast: Valuation of Jaguar (*Panthera onca*) Tourism and Cattle Depredation in the Brazilian Pantanal." *Global Ecology and Conservation* 11 (2017): 106–14.

Turner, Jack. *The Abstract Wild*. University of Arizona Press, 1996.

U.S Fish and Wildlife Service. "Endangered and Threatened Wildlife and Plants: Final Rule to Extend Endangered Status for the Jaguar in the United States." *Federal Register* 62, no. 140 (July 22, 1997): 39147–57.

———. "Endangered and Threatened Wildlife and Plants: Designation of Critical Habitat for Jaguar." *Federal Register* 79, no. 43 (March 5, 2014): 12571.

———. Letter to Michael Robinson, re: Petition to Reintroduce the Jaguar as an Experimental Population, January 10, 2024.

Upchurch, Marilyn. "Petition to Reintroduce Jaguars to Gila National Forest Denied by Federal Government." KRQE News13, February 5, 2024.

Valderrama-Vasquez, Carlos, et al. "Predator-Friendly Ranching, Use of Electric Fences, and Creole Cattle in the Colombian Savannas." *European Journal of Wildlife Research* 70, no. 1 (2023).

Vargas Soto, Juan S., et al. "Human Disturbance Shifts the Vertebrate Community from Large-Bodied Threatened Species to Small-Bodied Generalists in a Neotropical Biodiversity Hotspot." *Conservation Biology* 36, no. 2 (2022): e13813.

Veit, Peter, David Gibbs, and Katie Reytar. "Indigenous Forests Are Some of the Amazon's Last Carbon Sink." World Resources Institute, January 6, 2023.

Verheij, Pauline. "An Assessment of Wildlife Poaching and Trafficking in Bolivia and Suriname." International Union for Conservation of Nature, 2019.

Wagner, Dennis. "The Cat, the Captors and the Cover-Up." *Arizona Republic*, December 9, 2012.

———. "Mistakes Doomed Jaguar Snared in Mexico." *Arizona Republic*, December 11, 2012.

———. "Web of Intrigue Surrounds Death of Jaguar Macho B." *Arizona Republic*, December 9, 2012.

Walker, Matt. "Jaguar Mums Give Up Baby Secrets." *Earth News*, May 29, 2009.

Wallace, Scott. "An Illegal Gold Rush Is Igniting Attacks on Indigenous People in the Amazon." *National Geographic*, July 6, 2021.

Walsh, Bryan. "The Indiana Jones of Wildlife Protection." *Time*, January 10, 2008.

Wantzen, Karl M., et al. "The End of an Entire Biome? World's Largest Wetland, the Pantanal, Is Menaced by the Hidrovia Project Which Is Uncertain to Sustainably Support Large-Scale Navigation." *Science of the Total Environment* 908 (2024): 167751.

Warshall, Peter. "When Will Female Jaguars Cross the Border? Socio-Demographics of the Northern Jaguar." In *Merging Science and Management in a Rapidly Changing World: Biodiversity and Management of the Madrean Archipelago III and 7th Conference on Research and Resource Management in the Southwestern Deserts, 2012 May 1–5, Tucson, AZ*, ed. Gerald J. Gottfried et al., 87–90. U.S. Department of Agriculture, Forest Service, Rocky Mountain Research Station, 2013.

Webb, Elizabeth. *American Jaguar: Big Cats, Biogeography, and Human Borders*. Twenty-First Century Books, 2022.

Webber, Jeffery R. "Debunking Myths: The Eastern Lowlands of Santa Cruz." *AIN Bolivia*, January 2010.

Wei He, Lucía. "How China Is Closing the Soft Power Gap in Latin America." *Americas Quarterly*, April 12, 2019.

Wesdock, Nick. "Scat: The First Step in Conserving Mesoamerican Jaguars." *Wildlife Society*, November 11, 2016.

Wilcox, Sharon. "Encountering El Tigre: Jaguars, Knowledge, and Discourse in the Western World, 1492–1945." PhD diss., University of Texas at Austin, May 2014.

———. "Savage Jaguars, King Cats, and Ghostly Tigers: Affective Logics and Predatory Natures in Twentieth-Century American Nature Writing." *Professional Geographer* 69, no. 4 (2017): 531–38.

Wildlife Conservation Society. "Dramatic Photo Shows Jaguar and Cubs from Bolivia's Kaa Iya National Park" (press release). WCSNewsroom, December 21, 2011.

Wink, Georg. "Brazil, Land of the Past: The Ideological Roots of the New Right." *Bibliotopía*, 2021.

Wynne, Jut. "Belize's Maya Forest Corridor: A 'Missing Link' to Giant Rainforest Preserve." *Mongabay*, May 24, 2021.

Zeller, Katherine A. *Jaguars in the New Millennium Data Set Update: The State of the Jaguar in 2006*. New York: Wildlife Conservation Society, 2007.

Zeller, Katherine A., Alan Rabinowitz, et al. "The Jaguar Corridor Initiative:

A Range-Wide Conservation Strategy." In *Molecular Population Genetics, Evolutionary Biology, and Biological Conservation of Neotropical Carnivores*, ed. Manuel Ruiz-Garcia and Joseph M. Shostell. Nova Science Publishers, 2013.

Zeller, Katherine A., Sahil Nijhawan, et al. "Integrating Occupancy Modeling and Interview Data for Corridor Identification: A Case Study for Jaguars in Nicaragua." *Biological Conservation* 144, no. 2 (2011): 892–901.

Index

Abbey, Edward, 82
Acatlán, 50
Acosta, Luis, 118
Acurizal Farm, 24, 25
Acurizal Reserve, 152
Africa, 39, 233, 235–36
Ah Puch (jaguar), 33, 62–65
Alanen, Marit, 258
"Alan Rabinowitz; Indiana Jones Meets His Match in Burma" (Line), 100
Alan Rabinowitz Research Center, *see* Dr. Alan Rabinowitz Research Center
Aldo Leopold wilderness, 255
Alexander, Jane, 57, 95, 97, 99, 101, 130, 150, 153
Almeida, Tony de (Antônio Eduardo d'Andrada Almeida), 25
alpha predators, 40
Altiplano settlers, 110
Amazon forest, 166–72
Amazon River, 149
Ambushed on the Jaguar Trail (Childs), 199
American Museum of Natural History, 151
American Serengeti (Flores), 13
American Society of Mammalogists, 14
Amistad, La (Costa Rica and Panama), 218
AMLO, *See* Obrador, Andrés Manuel López
Ampara Animal Institute, 173
anacondas, 153
Angier, Natalie, 47, 76
animal massacres, 13
animal parts trade, 100, 116–18, 156, 235; *See also* jaguar(s)
Animal Research and Conservation Center, 27
Animas, N. Mex., 202

Animas Mountains (New Mexico), 203
Annamite Mountains, 98
anteaters, 108–9, 143, 153
anti-predation methods, 113
Apache-Sitgreaves National Forests, 2
Apex Silver, 140
Arawak-speaking people, 48
Argentina, 259
Arias, Melissa, 230–31
Arizona; *See also* Central Arizona-New Mexico Recovery Area; New Mexico and Arizona Cattle Growers' Association; University of Arizona; specific places, e.g.: Chiricahua Mountains
  Border wall in, 245, 261
  El Jefe in, 1–9
  groundwater regulation in, 247
  jaguar populations in, 204–5
  predator status of jaguars in, 13–14
Arizona Capitol Times, 209
Arizona Daily Star, 3, 208, 210, 247, 251
Arizona Deer Association, 266
Arizona Ecological Services, 209
Arizona Game and Fish Department (AZGFD), 3, 7, 200–201, 206–10, 213–14, 258, 266
Arizona–New Mexico Jaguar Conservation Team (JagCT), 201–3, 205, 207, 210–11
Arizona Republic, The, 214
Arizona State Commission, 199
Arizona State University, 171
Arizona Trail Association, 253
"Arizona Trail Cam Journey, An" (Miller), 265
Assembly of Defenders of Maya Territory (Múuch' Xíinbal), 227

Atascosa Mountains (Arizona), 205
Atlas of a Lost World (Childs), 41
"At Last, Ready for Its Close-Up" (Angier), 47
Audubon, John James, 13, 24
Austin, Josiah, 242
Avila, Sergio, 211, 259
AZGFD, *See* Arizona Game and Fish Department
Azofeifa, Alejandro "Champeon," 73–74, 77–81, 91–94
Aztec civilization, 52–53

Babb, R. D., 2
Baboquivari Mountains (Arizona), 197–98
Balme, Guy, 150
Barrio Brewing Company, 8
bats, 115, 159
bears, 181, 244
Beautiful, The (jaguar), *See* El Bonito (jaguar)
beavers, prehistoric, 41
beef cattle, 142, 146
beehives, 220
Belen outdoor market, 19
Belize, 2, 17, 67–68; *See also* Cockscomb Basin (Belize)
Belize Audubon Society, 69–70
Belt and Road program, 117, 232
Bennett, Elizabeth, 236
Bering Land Bridge, 38–41, 268
Berton, Eduardo Franco, 19–21
Beyond the Last Village (Rabinowitz), xiii
Biological Conservation, 20
BioScience, 251
birds, 125, 129, 143, 159, 183, 249
bison, prehistoric, 41
black bears, 54
bobcats, 244
Bolivia; *See also* Madidi National Park; Mojeño people
  fires, 122–23
  habitat preservation efforts, 107–12, 114–15, 126–27
  jaguar parts trade, 19, 21, 163, 229–30, 234–36
  and poaching, 115–18
Bolivia-Brazil gas pipeline, 122–24
Bolivian National Police, 117
Bolsonaro, Jair, 152, 154, 168–69
Borderland Jaguars (Brown and López-González), 49
Borderlands Jaguar Detection Project, 202, 213, 214

Borderlands Secondary (Jaguar) Area, 253, 256
Border Patrol, 248, 250
Border Wall, U.S.-Mexico, 221, 245, 248–52, 257, 261, 267
Borneo, 70–71, 98
Boron, Valeria, 162–64
Boston, Mike, 81
Boulhosa, Ricardo, 142
Brazilian Amazon, 15
Brazilian Pantanal, xv, 12, 16
British Honduras, *See* Belize
Bronx Zoo, 35, 151
Brown, David, 49
Brun, Janay, 208, 213, 215
Bugbee, Chris, 4, 6–9, 214, 251, 261
Bureau of Land Management, 238
Burma, *See* Myanmar

CABI (Capitanía del Alto y Bajo Izozog), 121
Caiman Ecological Refuge, 144
Caiman Ranch, 192
caimans, *See* crocodiles
Cajón Bonito, Mexico, 238–39, 242–43, 245
Calakmul, 216–19, 221, 225
Calakmul Biosphere Reserve, 217
Calakmul Jaguar Conservation Unit, 223
Cambridge University, 122
camera traps, 87, 91–92, 115, 220, 254
Campeche, Mexico, 219–20
CANRA (Central Arizona-New Mexico Recovery Area), 255–56
Capitanía del Alto y Bajo Izozog (CABI), 121
capture and collaring, of jaguars, 181–83, 188–96, 203–4, 206–9
capybaras, 143, 150
caracaras, 163
caribou, prehistoric, 41
Carr, Archie, Jr., 27
Carr, Archie "Chuck," III, 27, 34, 60–62, 68, 105
Carrillo, Eduardo, 73, 75–76, 80, 87, 160, 204, 263
Carson, Rachel, xv
Caso, Alfonso, 49
Cassini, Oleg, 14–15
catamounts, *See* pumas
catfish, 153
cattle and cattle ranches, 142–47, 153, 155
CBS This Morning (TV show), 8
Ceballos, Gerardo, 217, 226, 227, 239

Center for Biological Diversity
  and copper mining, 246–47
  and critical habitat lands, 257–58, 261, 266
  lawsuits and investigations, 200, 202, 209, 211, 213
  publicity campaign, 5–9
Centla Biosphere Reserve, 220
Central Arizona-New Mexico Recovery Area, 255
Centro Universitario Univinte, 150
Chac (jaguar), 58–59
Chaco chachalacas, 159, 161, 180
Chasin' Tail Guide Service, 2
Chavín civilization, 51–52
Chichén Itzá, 52
Childs, Anna Mary, 198–202
Childs, Craig, 41
Childs, Jack, 198–202, 204–5, 214–15
China, 19, 21, 55, 117–18, 153–54, 228–33, 235–36
"China's Lust for Jaguar Fangs Imperils Big Cats" (Fraser), 231
Chiquitano forest (Bolivia), 111, 115
Chiricahua Apaches, 3
Chiricahua Mountains, 237–38, 256
Chor, Laurel, 119
Cirillo (Maya tracker), 59, 69
CITES, *See* Convention on International Trade in Endangered Species of Wild Fauna and Flora
civilizations, pre-Colombian, 48–53; *See also* Maya and Mayan people
Claiborne, Liz, 102
Clark, Valer, 240–42, 254
climate change, 170–71
Cline, Michael, 264
Clinton, Bill, 57
Cloak and Jaguar (Brun), 215
clouded leopards, 44, 70–71, 95, 98, 235
coatis, 174
cocaine, 235
Cochise (Apache leader), 3
Cochise (jaguar), 265
Cockscomb Basin (Belize), 23, 26–27, 30 34, 58–62, 65–70, 99, 103
Cockscomb Forest Reserve/Jaguar Preserve, 69
Coe, Michael, 50
Colombia, 11, 48, 163, 214, 218
Columbia University, 151
Coming into the Country (McPhee), 82
Conservación Loros Bolivia, 109

Conservation Biology, 132, 236
Conservation CATalyst, 9
Conservation Land Trust, 259
Conservation Planning Specialist Group, 258
Conservation Science and Practice, 255
Convention on International Trade in Endangered Species of Wild Fauna and Flora (CITES), 15–17, 22, 156, 228–30, 232, 234, 265
Convention on Migratory Species, 265
Cooperative Fish and Wildlife Unit (University of Arizona), 210
Copper World Complex, 246–47
Coral Reef National Monument, 176
Corcovado National Park (Costa Rica), 73–77, 80–83, 85–91, 160
Corixo Negro, Brazil, 173
Cornell University, 127, 151
Coronado National Forest, 256
corridors, *See* Jaguar Corridor Initiative; Mexican and Central American Jaguar Corridor
Corumbá de Goiás, 174
Costa Rica, 75, 133–34, 218; *See also* Corcovado National Park
cougars, 181
COVID-19 pandemic, 86, 126–27, 151–52, 184–85, 218, 221
Crawshaw, Peter, 24–25, 55, 103, 136, 185, 202
crocodiles, 153, 156–57, 174
Crook, George R., 3
Crosta, Andrea, 233–36
Crowder, Clay, 258
Cruz, Juan Carlos, 86
crypsis, 75
Cuenca Los Ojos, 237, 240, 242
Cuiabá (city), 154–55
Cuiabá River, 145, 148, 149, 152, 172
Culver, Melanie, 203, 238
Cuvier, Georges, 21
cyanide bombs, 13

da Costa, Jamil Rodrigues, 183–84
dams, loose-rock (trincheras), 241–42
Dani (ranger), 79–80, 89–91, 93–94
Darién Gap, 137, 218, 263
Darwin, Charles, 21, 24–25, 43
data and data collection, 59, 87, 103–4, 132–33, 175, 181–82, 229
Davis, Tony, 208, 210, 212
deer, 153, 244

Deer and the Tiger, The (Schaller), 71
Defenders of Wildlife, 213
Defense Department, 249
deforestation, 120, 166–72
de Gaulle, Charles, 14
Department of Homeland Security, 207, 215
Department of Justice, 210
Desana people, 48
Devlin, Allison, 150–52, 157, 165, 189–91, 193–95, 267
DiCaprio, Leonardo, 169
Di Martino, Sebastián, 259
Dito (ranch hand), 192
Dolmatoff, Gerardo Reichel, 48
Dominican Republic, 133
Dos Cabezas Mountains (Arizona), 198, 238
Douglas, AZ, 237–38
Dr. Alan Rabinowitz Research Center, 263–64
ducks, 165
Duguid, Julian, 152
Duke University, 219
Dzib, Eleazar, 225

Eagle Knights, 53
Earth League International (ELI), 233, 235
eco-tourism, 126–27, 147
Ecuador, 234
education, 144
egrets, 159, 163, 164
Eizirik, Eduardo, 104
El Bonito "The Beautiful" (jaguar), 237, 239, 243, 245
El Coronado Ranch, 241–42
electric fences, 147
Electrum Group, 141
elephant, 235–36
Elephant Action League, 233
ELI (Earth League International), 233, 235
El Jefe (jaguar), 1–9, 18, 246, 251–52
elk, prehistoric, 41
El Salvador, xv
empty forest syndrome, 83, 99
Encontro das Águas State Park, 173
Endangered Species Act (ESA), 5, 6, 200, 201, 212, 256–57
Endangered Species Center (San Diego Zoo), 210
Endangered Species Conservation Act (ESCA), 199–200
Endangered Species Preservation Act, 199

Environmental Impact Manifestation (MIA), 226
Eradication Methods Laboratory, 13
ESA, *See* Endangered Species Act
ESCA (Endangered Species Conservation Act), 199–200
Estância Miranda, 25
European arrival, 12
"Evolution and Genetics of Jaguar Populations" (Eizirik), 103–4
extinctions, of prehistoric animals, 41–42
Ezra (research assistant), 106, 108, 115, 122–23, 125, 128–29

Fauna VIVA, 116
Fawcett, Percy, 152
Federal University of Mato Grosso, 174
Fenn, Alyson, 2–3
Fenn, Donnie, 2–3
fer-de-lance snake, 81–82
Figel, Joe, 167
Figueirôa, Gustavo, 168
Figueroa, Omar, 70
fires, 110–11, 122–23, 166–75, 178
Fisher, Kim, 256–57, 261
Five Great Forests of Mesoamerica initiative, 218
Flesch, Aaron, 251
flood pulse, 149
Flores, Dan, 13
Fonatur (National Fund for Tourism Development), 226
Forbes magazine, 145
forest fragmentation, 220
Forest Service, 6, 7
Fort Apache Indian Reservation, 2
Fossey, Dian, 72
fossils, 39
Foster, Rebecca, 151
Four Jaguar (Toltec leader), 52–53
foxes, 153
France-Soir, 14
Fraser, Barbara, 231
Freedom of Information Act, 208
Friedeberg, Diana, 219–20, 225
frogs, 128–29
Fuchs, Richard, 154
Full Cry (magazine), 199
FUNAI (the National Indian Foundation), 168
Fundación Rewilding Argentina, 259

Gadsden Hotel (Douglas, AZ), 238
Ganesh Marín, Alejandro, 237–45
Garcia, Nabhan, 166–67
Garfield, Ezra, 183–84, 190
gas pipeline, Bolivia-Brazil, 122–24
Geronimo (Apache leader), 3
Gessner, David, 254–55, 260
Gila National Forest (New Mexico), 255, 257, 258
Glenn, Warner, 197–98, 201–3, 249
Global Forest Watch, 220
gold, 83, 86
Goldman, E. A., 14
Golfo Dulce Forest Reserve, 89
Goodall, Jane, xiv–xv, 72, 234
Good Morning America (TV show), 8
GRAD (Group for Animal Rescue in Disasters), 173–74
Gran Calakmul region, 224
Gran Chaco region, 120–21, 126
Grand Canyon, 255
Gran Iberá National Park, 259
Great Acceleration, xv
Great Jaguar (Ursa Major) constellation, 53
Great Vasyugan Mire (Siberia), 171
Green Hell (Duguid), 152
"Green Lagoons, The" (Leopold), 84, 260
Grinnell, Joseph, 14
Group for Animal Rescue in Disasters (GRAD), 173–74
Guahibo people, of Venezuela, 48
Guanacaste National Park (Costa Rica), 263
Guaraní people, 121
Guardian, The, 110
Guatemalan Highlands, 52
Guermo (dog handler), 29–30, 33
Guerra de la Sed, La (The War of Thirst), 121
Guidelines for Species Conservation Planning (IUCN), 258
Gulf of Mexico, 220
Guyana, 234

Haaland, Deb, 257
Haberfeld, Mario, 144
Harmsen, Bart, 151, 264
Hassan, Bader, 28–33, 60
Haub School of Environment and Natural Resources, 239
hawks, 163
Heffelfinger, Jim, 258, 266
herons, 159, 163

Hiaki High School (Tucson, AZ), 240
"High Jaguar Densities" (Tobler, Carrillo-Percastegui), 20
Higuero, Ivonne, 228
Hijacking My Dream (Larsen), 114
Holzmann, Anai, 108, 112–14, 116–17, 119
Honduras, 134, 218
Hoogesteijn, Rafael, 12, 146–47, 161
Hovatter, Gary, 214
Htamanthi Wildlife Sanctuary, 99
Huachuca Mountains, 240, 254
Huai Kha Khaeng Wildlife Sanctuary, 70–71, 235
Hudbay Minerals, 247
Hukawng Valley, 101
human migration to Americas, 39–40
humans, early, 41–42
Humboldt, Alexander von, 21, 24–25
Humboldt University, 111
Hunter, Luke, 142, 145, 183
Hunter, Malcolm L., Jr., 239
hydrocarbons, 140

ibis, 159
Ihering, Rodolpho von, 25
"Illegal Jaguar Trade in Latin America" (Arias), 230
Illegal Trade in Jaguars, The (CITES), 228
Indigenous Territories (ITs; Brazil), 167–68
Indio Maíz-Tortuguero (Nicaragua and Costa Rica), 218
Indomitable Beast, An (Rabinowitz), 84, 104, 106, 131, 213
Institute of Ecology, 217
Institute of Maya Studies, 217
Institute of Meteorology and Climate Research, 154
Institutional Animal Care and Use Committee, 182
Integrated Surveillance Towers, 248–49
International Fund for Animal Welfare, 234
International Union for the Conservation of Nature (IUCN), xv, 234, 258
"In the Front Line of the Cold War" (Kaplan), 139
Irish Examiner, 231
Isoseño-Guaraní people, 121–22
Isoso Camp, 124–25, 127
ITs (Indigenous Territories; Brazil), 167–68
Itzamná (jaguar), 59

IUCN (International Union for the Conservation of Nature), xv, 234, 258
Ixchel (jaguar), 60–61

jacanas, 163
JagCT, *See* Arizona–New Mexico Jaguar Conservation Team
jaguar(s)
  anatomy of, 44–46
  census, 217
  habitat, 210–13, 231; *See also* forest fragmentation
  as hunters, 44–47, 186
  hunters and hunting of, 2, 12–21, 25, 77, 155–56, 197, 231
  migration to Americas, 38–43
  in myth, 11–12, 48–53, 76
  parts trade, 18–22, 108, 119, 163, 185, 228–32, 234–36
  pelts, 15–17
  populations, 13–15, 126, 204
  range and conservation status, xv, 15, 103
  sedation of, 31–33
  taxonomy, 103–5
  as term, 74–75
  as threat to livestock, 147
Jaguar, The (Hoogesteijn and Mondolfi), 161
Jaguar Cars of Canada, 70, 103
Jaguar: Chasing the Dragon's Tail (Rabinowitz), xiii, 30, 34, 82
Jaguar Claw (Mayan ruler), 216
Jaguar Conservation Trust, 103
Jaguar Conservation Units (JCUs), 66, 126, 135–36, 141, 167, 267
Jaguar Corridor, 132–38
  establishment of, 263
  hunting in, 14, 18
  and Jaguar Conservation Units, 167
  and Journey of the Jaguar, 107
  and Maya Train, 221
Jaguar Corridor Initiative, 16, 66, 172, 265
Jaguar Cultural Corridor, 49, 102, 138
Jaguar Hunting in the Mato Grosso and Bolivia (Almeida), 25
Jaguar in the New Millenium, The, 103
Jaguar Knights, 53
Jaguar: One Man's Struggle to Establish the World's First Jaguar Preserve (Rabinowitz), xiii, 34, 61, 96–97, 133, 264
JaguarOsa Project, 79, 87–88
Jaguar Protector (jaguar), 254
"Jaguar Protector" (jaguar), 254

Jaguar Recovery Team (JRT), 256
Jaguars in New Millennium (Zeller), 133
Jaguar's Path, The, 113
Jaguar Survey and Monitoring Project, 3, 7
jaguarundis, 129
Jason (ranger), 81–82, 85, 90–91
JCUs, *See* Jaguar Conservation Units
Jelinek, Arthur, 42
Jesus birds, 163
Jiménez, Ernesto Martínez, 226
Jofre Velho Ranch, 145–46, 148–50, 172, 181
Johnson, Terry, 201, 207, 210
Jorge (jaguar), 175–77
José (driver), 106, 108
Journey of the Jaguar (JotJ), 107, 220–21, 263
JRT (Jaguar Recovery Team), 256

Kaa-Iya del Gran Chaco National Park, 108, 120–22
Kaaiyana (jaguar), 124
Kaplan, Tom, 139–46, 264
Karlsruhe Institute of Technology, 154
Kempff, Yandery, 116, 118
Kennaugh, Alexandra, 232
Kennedy, Jacqueline, 14
Kennerknecht, Sebastian, 183–84
Kimbro, Kelley 197–98
Kipling, Rudyard, 99
Kogi people, of Colombia, 48
Koprowski, John, 239
Kyra (jaguar), 175–77

La Amistad International "Friendship" Park, 90
Laboratory of Genomic Diversity, 104
Laguna de Términos Conservation Unit, 219–20, 223
Lan, Yin, 117–19
Larmer, Brook, 231–32
Larsen, Duston, 108, 112–16
Larsen, Ronald, 114, 116
Larson, Shawn, 103–4
Leakey, Louis, 42
Leclerc, Georges-Louis, 21
leishmania, 83
leopard coats, 15
leopards, clouded, 44, 70–71, 95, 98, 235
Leopold, Aldo, 12, 14, 83–84, 260; *See also* Aldo Leopold wilderness
Leopold, A. Starker, 12, 45, 54, 160, 243

Leopold, Carl, 260
Life in the Valley of Death (Rabinowitz), xiv
Linné, Carl von, 44
Lives of Game Animals (Seton), 24
Living with Cats (Viviendo con Felinos), 18
Lopez, Barry, 13
López-González, Carlos, 49
Lordsburg, N. Mex., 200
Los Angeles Times, 96, 209
Los Ojos Ranch, 237
Luke Hunter, 145

MacArthur, R. H., 135
macaws, 58, 180
Macho A (jaguar), 202
Macho B (jaguar), 6, 204–15
"Macho B: Last Roar of the Jaguar" (Wagner), 214
Macho Uno (jaguar), 74, 88–89
Madidi National Park (Bolivia), 109, 119–20
Madison, Wisc., 210
Malinalco temple, 53
mammoths, prehistoric, 41
man-eating animals, 160–61
maned wolves, 120
Manocherian, Greg, 264
margays, 65
Martin, Abbie, 175–79
Martin, Paul, 42
Matthiessen, Peter, 71, 129
Matto Grosso, 149, 154
May, Joares, 150, 152, 162, 182–83, 188–95
Maya and Mayan people; *See also* Maya Train; Selva Maya
  civilization, ancient, 26–27, 50–51, 57–59, 216–18, 226
  mythology, 12
  present-day, 34, 69–70, 225, 227
Maya Cosmos (Freidel, Schele, Joy), 49
Maya Train, 221–22, 224–27
Mayke (dog), 4, 7–8, 251
McCain, Emil, 204, 206, 208, 213, 215
McCoy, Terrence, 168
McDaniel College, 35
McDowell, Don, Sr., 266
McFarland, Duncan, 151
McFarland, Ellen, 151
McPhee, John, 82
McPhee, Nick, 106–10, 122
McSpadden, Russ, 5, 8, 246–47
MeatEater (website), 266

Memoranda of Understanding (MOUs), 136–37
Menezes, Agamenon da Silva, 166
Mennonite people, 121, 220
Mennonites, 112
Mesoamerican Biological Corridor, 105
Mexican and Central American Jaguar Corridor, 214
Mexico, 11, 133, 217–19, 242; *See also* U.S.-Mexico border
  Viviendo con Felinos, 18
Mexico (Coe), 50
Mexico City, 53, 221
MIA (Environmental Impact Manifestation), 226
Miguel Colorado, 223
Miles, Nelson A., 3
Miller, Brian, 203
Miller, Jason, 265–66
Miller, Phil, 258
Miller, Tom, 64, 96
Miller Peak Trail Junction, 254
Ming, Li, 117–19
mining, 246–47
Mogollon Plateau, 255
Mogollon Rim, 2
Mojeño people, of Bolivia, 48
Mongabay (publication), 18–19
monkeys, 174
Monster of God (Quammen), 40, 160
Montana, 13
Montana Fish, Wildlife, and Parks, 257
Montana State University, 122
Montezuma Pass Trailhead, 253
Montreux Record, 169
moose, prehistoric, 41
Morales, Evo, 110–11, 114
Morcatty, Thaís, 232
Moskitia, La (Nicaragua and Honduras), 218
mountain lions, *See* pumas
Mount Hkakabo Razi, 99
Mourão, Hamilton, 168
MOUs (Memoranda of Understanding), 136–37
Mule Deer Working Group, 266
Murie, Olaus, 14
Murillo, Juan Carlos Vasquez, 229
musk oxen, prehistoric, 41
Múuch' Xíinbal (Assembly of Defenders of Maya Territory), 227
Myanmar, 96, 98, 100–102, 119, 131

Nahui Ocelotl (Four Jaguar; Toltec leader), 52–53
NAMA Conservation, 86–87
Nash, Jack, 140
National Alliance for Jaguar Conservation, 217
National Forestry and Wildlife Service, Peru (SERFOR), 20
National Fund for Tourism Development (Fonatur), 226
National Geographic, 237
National Indian Foundation (FUNAI), 168
National Institute of Biodiversity, 77
National Park System Act, 66
National System of Conservation Areas (SINAC), 88
Natural History Museum (London), 103
Naturalia, 18, 240
Naturalist and Other Beasts, A (Schaller), 36, 45, 185
Natural Protected Areas system (Mexico), 242
Nature (journal), 42
"Nature Divided, Scientists United" (Peters), 251
Navia, Roberto, 21
necropsies, 210
Neils, Aletris, 7, 8
Nelson, Matt, 253
Ñembi Guasu Area of Conservation and Ecological Importance, 126
New Mexico, 13–14; *See also* Arizona–New Mexico Jaguar Conservation Team (JagCT); Arizona–New Mexico Jaguar Conservation Team (JagCT); Central Arizona–New Mexico Recovery Area
Border Wall in, 245
New Mexico and Arizona Cattle Growers' Association, 211
New Mexico Cattle Growers Association, 266
New Mexico Department of Agriculture, 211
New Mexico Department of Game and Fish, 200, 210
New Mexico Farm and Livestock Bureau, 257
New York Times, 47, 76, 100, 114, 170
New York Zoological Society, 16, 17, 27, 34, 55, 68; *See also* Wildlife Conservation Society
NEX (No Extinction) Institute, 174
Nicaragua, 134, 218
Nicte (jaguar), 219–20
Nijman, Vincent, 231
Nimer, Stephen, 130

Nixon, Richard, and administration, 199
Noel Kempff Natural History Museum, 112
No Extinction (NEX) Institute, 174
Northern Jaguar Project, 240
Northern Jaguar Reserve (Mexico), 1, 240
Northland College, 82, 87
Northwestern [Jaguar] Recovery Unit (NRU), 256–57
Nyunt, Khin, 131

Obama, Barack, and administration, 211
Obrador, Andrés Manuel López "AMLO," 221, 224, 227
ocelots, 58, 65, 120, 143, 150, 162
Old Colony Mennonites, 112
Old Crow Basin (Yukon Territories), 39
Olmec civilization, 49–50, 52
Olson, Erik, 82, 87–88
On Being (radio show), xiv
Oncafari Association, 144
Operation Jaguar, 234
Origin of Species, On the (Darwin), 24, 43
Ortenberg, Art, 102–3
Oryx (journal), 255
O:ṣhad Ñu:kudam ("Jaguar Protector") (jaguar), 254
otters, 143, 153
Otuquis National Park and Integrated Management Natural Area, 126
Ousado (jaguar), 173–75, 181
owls, 260–61
Oxford Brookes University, 231
Oxford University, 139, 144, 230

Pacific Tropical Wet Forest, 74
Panama, 218
pandas, 55, 181
Pantanal, 24–26, 55–56, 136, 142–46, 148–58, 169–72, 175–78, 180–87
Pantanal Conservation Area, 157
Pantanal Jaguar Identification Project, 175–76
Pantanal National Park (Brazil), 152
Pantaneiro cattle, 147
Panthera (organization), 12, 76, 90, 113, 140–47, 219, 264–65, 267
Panthera ranch, 152
panthers, *See* pumas
Paraguay, 120–22
Paraguay River, 149, 171
parakeets, 165
Parapetí River, 121
Patagonia Mountains (Arizona), 2

Paterson, Tom, 266
Patricia (jaguar), 175–79, 186
Peacock, Doug, 82
peccaries, 55, 80–81, 87, 192
Peloncillo Mountains (New Mexico), 197
Perez, Pedro, 20
Perry, Richard, 25, 49
Peru, 19, 234
Peruvian Amazon Research Institute, 20
Peters, Rob, 251
Petrobras, 122
Phoenix Zoo, 209
Piedras Blancas National Park, 90
Piran, Adecio, 166
Pisaro, Andrea, 162–63
platinum, 140
Pleistocene era, 38–41, 43, 45
Pleistocene extinctions, 12
plovers, 159
poaching, 86, 95, 119, 231, 235–36, 240; See also animal parts trade
Pocock, Reginald, 103–4
Poconé, 148
Polisar, John, 16, 229, 266–67
Pop, Ignacio, 99–100
Porfilio (Maya assistant), 32–33
Porto Jofre, 141, 148, 181
Povilitis, Tony, 200, 205, 211, 261
precolonial people, 11–12
pre-Colombian civilizations, 48–53
predator control, 13–14
Price, George Cadle, 68–69
pronghorn, prehistoric, 41
"Proposal for a Conservation Inventory of Threatened Species" (Leopold), 84
pumas, 4, 58, 65, 76, 124–25, 143, 243–44
Pyne, Stephen, 168

Quammen, David, xv, 40, 160
Quigley, Howard, xiv, 34, 54–57, 144–47, 219–22, 263

Rabinowitz, Alan; *See also* Dr. Alan Rabinowitz Research Center; Rabinowitz, Salisa Sathapanawath
 among Mayan people, 58–64
 in Belize, 26–30
 Beyond the Last Village, xiii
 in Calakmul, 223–24
 and cattle ranches, 142–45
 and Childs and Glenn families, 201
 and Cockscomb Region, 65–72
 and death of Macho B, 212–14
 and Allison Devlin, 150–51
 early life, 34–36
 and Diana Friedberg, 219
 and Guermo (jaguar), 31–33
 and Hukawng Valley, 130–31
 illness and death, 101, 130, 221, 263–64
 Indomitable Beast, An, 35, 84, 104, 106, 131, 213
 Jaguar: Chasing the Dragon's Tail, xiii, 30, 34, 82
 and Jaguar Corridor, 132–39
 Jaguar: One Man's Struggle to Establish the World's First Jaguar Preserve, xiii, 34, 61, 96–97, 133, 264
 on jaguars as commodities, 22
 "Jaguars in the New Millennium" conference, 102–5
 and Tom Kaplan, 139–45, 147
 Life in the Valley of Death, xiv
 mapping of Jaguar Corridor, 214
 marriage, 95–97
 in Myanmar, 98–102
 and Khin Nyunt, 131
 and Howard Quigley, 54–57
 in Thailand, 236
Rabinowitz, Alana Jane, 130, 263–64
Rabinowitz, Alexander, 130, 263–64
Rabinowitz, Frank "Red," 35
Rabinowitz, Salisa Sathapanawath, 96–98, 101, 263–64
radio collars, 32–33
Radwin, Maxwell, 87
Ragan, Kinley, 243
Ramsar Convention on Wetlands, 169
Rancho San Bernardino, 242
range fidelity, 124
Recanati, Daphne, 140
Recanati, Leon, 140
Reed, Jim, 217
reindeer, prehistoric, 41
Repenning, Charles, 38
rewilding, 253–61
Reyes, Casandra, 226
rhinos, 235–36
Río Bavispe, 238
Río Yaqui, 238
riparian areas, 249
Riparian National Conservation Area, 250
ritual sacrifice, 51
Rivero-Guzman, Kathia, 118
Robinson, Michael, 203, 211–12, 257, 261, 267

Rodríguez-Echandi, Carlos Manuel, 133–34
Roll, John, 210, 211
Romero-Muñoz, Alfredo, 111
Roosevelt, Kermit, 155
Roosevelt, Theodore, 12, 155
Rosemont (copper mine), 246–47
Royal Forest Department (Thailand), 235
Rumiz, Damián, 112, 115–19
Ruta del Jaguar, La (The Jaguar's Path), 113

Sabin, Andy, 57, 264
Saborío, Guido, 88
Salles, Ricardo, 168–69
Salomão, Jorge, 174
Salom-Pérez, Roberto, 76, 80, 90, 134–35
San Carlos Apaches, 256
Sanderson, Eric, 132–33, 256–58, 261
San Diego Natural History Museum, 209
San Diego Zoo, 210
sandpipers, 125
San Francisco Peaks (Arizona), 256
San José de Chiquitos, 120
San Miguelito Jaguar Conservation Ranch, 111–14
San Pedro River, 249–50
Santa Cruz de la Sierra, Bolivia, 109, 111
Santa Rita Mountains (Arizona), 2–6, 246–47, 260
Santa Rosa National Park, 89
Santiago Alonso, Carmen, 221
Santos, Eurico, 25
Santos, Francisco, 136, 214
Sassi, Carla, 174
Sathapanawath, Salisa, *see* Rabinowitz, Salisa Sathapanawath
Saul (biologist-guide), 122–28
Saunders, Nicholas, 49, 53
Schaller, George, 16–17, 23–28, 36, 54–55, 70–71, 129, 144–47
Schaller, Kay, 55
sedation, of animals, 31–33, 174, 183, 193, 208
Selva Maya, 217–18
Sepulvida, Raíssa, 162–63, 191, 193–95
Serengeti Lion, The (Schaller), 37
SERFOR (National Forestry and Wildlife Service, Peru), 20
Sergio (Indigenous worker), 224
Serraglio, Randy, 5, 8, 246–47, 252–54, 260
Seton, Ernest Thompson, 24
Seymour, Kevin, 45
shamans, 48

sheep, prehistoric, 41
Sierra Club, 259
Sierra Leone, 162
Sierra Madre Mountains, 1
Sierra Madre Occidental, 238
Sikuani people, 48
silver, 140
Simon, Bob, 183
Simpson, Dairen, 150, 188
SINAC (National System of Conservation Areas), 88
Sirena Ranger Station, 74, 89
60 Minutes (TV show), 183
Skeat, Walter William, 75
skunks, 244
Sky Island Alliance, 211
Sky Islands (Arizona), 198, 250, 254–55, 258
Smith, Chad, 257
Smoky Mountains, 36
snakes, 28, 29, 81–82, 85, 118, 127; *See also* specific snakes
Snow Leopard, The (Matthiessen), 129
Sofia (jaguar), 194
Sombra (jaguar), 265
Sonora, Mexico, 5, 18, 237–40, 242
Soros, George, 140
Soros, Paul, 140
South Africa, 140
soybeans, 154
soy crops, 111–12
Spangle, Steve, 201, 209
Spencer, Glenn, 248
spoonbills, 159
"spreading the skin," 50–51
Stahl, Johannes, 229
Stark, Mike, 5
Starnes, Warren "Bud," 211
Steller, Tim, 208, 211–12
stone transport, 50
storks, 153, 163
Stuart, James, 210
SUNY College of Environmental Science and Forestry, 151
Superstition Mountains, 256
Suriname, 48, 163, 234
Switzerland, 169

Talamanca-Central Volcanic Region, 134
Talamanca Mountains, 90
tapirs, 120, 143, 150, 153, 174
tariffs, 153
tayras, 129

TCUs (Tiger Conservation Units), 66
Tenochtitlán, 53, 221
Teotihuacán, 52
Teton Cougar Project, 145
Texas, 140
Thailand, 71–72, 95–96, 235
Theory of Island Biogeography, The (Wilson and MacArthur), 134
Through the Brazilian Wilderness (Roosevelt), 155
ticks, 92
tiger(s), 100, 145
Tiger Conservation Units (TCUs), 66
Tiomkin, Avi, 140
Tippett, Krista, xiv
Tobler, Mathias, 20, 167
Today Show (TV show), 8
Toltec civilization, 52–53
Tompkins, Doug, 140, 259
Tompkins, Kristine, 140, 259
Tortato, Fernando
   and data collection, 181
   and fires, 170–72
   and habitat loss, 153–54
   and jaguar capture, 189, 191, 193, 195
   and jaguar tourism, 147
   and Jofre Velho Ranch, 149–50
   and Ousado (jaguar), 175
Tortuguero National Park, 89
tourists and tourism, 164–65, 181, 184; *See also* Maya Train
Tracking the Felids of the Borderlands (Childs), 202
Transpantaneira Highway, 148
Traphagen, Myles, 261
Travels and Research (Humboldt), 24
Tren Maya, *See* Maya Train
Tres Fronteras, 20
Três Irmãos River, 173, 177, 183
trincheras (loose-rock dams), 241–42
Trump, Donald, and administration, 117, 153, 245, 249; *See also* Border Wall, U.S.-Mexico
Tsimane mythology, 12
Turner, Ted, 140
2030 Roadmap, 265

Uc, Pedro, 227
Uexküll, Jakob von, 46
Ugalde, Álvaro, 75, 77
Umwelt (term), 46
Unger, Elizabeth, 119

United Nations Development Program, 97, 265
United Nations Environment Program, 265
United Nations Office on Drugs and Crime, 265
Universidad del Pacífico (Lima, Peru), 127
Universidad Nacional Autónoma de México, 102, 217
University of Arizona, 4, 6–7, 42, 210, 215, 239
University of California at Davis, 210
University of California at Santa Cruz, 200
University of Costa Rica, 76
University of Oxford, 230
University of Wyoming, 239
"Updates of Historic and Contemporary Records" (Babb), 2
Ursa Major (Great Jaguar) constellation, 53
Uruguay, xv, 22
U.S. Attorney's Office, 214
U.S. Bureau of Biological Survey, 13–14
U.S. Fish and Wildlife Service (USFWS), 7, 200–203, 208–10, 212, 214, 256–58, 266–67
USGS Wildlife Health Center (Madison, Wisc.), 210
U.S.-Mexico border, 1, 199, 204; *See also* Border Wall, U.S.-Mexico
Uzquiano, Marcos, 119–20

vaccination, of cattle, 147
Valencia Middle School (Tucson, Arizona), 5–6
Valerio (jaguar), 245
Van Pelt, William, 210
Venezuela, 48
Venta, La (Mexico), 49
Virgin Islands National Park, 176
Viviendo con Felinos (Living with Cats), 18
Viviparous Quadrupeds of North America, The (Audubon), 24

Wagner, Dennis, 214
Walker, Susan, 59–60, 62–63
Walkosak, Shiloh, 203
Wallace, Alfred Russel, 42, 160
Walsh, Brian, xiv
War of Thirst (Guerra de la Sed), 121
Warshall, Peter, 258
Washington Examiner, 239
Washington Post, The, 96, 168
Wasser, Samuel, 203

water buffaloes, 116, 147, 153
water conservation, 241–42
Wenwan (Chinese subculture), 231
Western Association of Fish and Wildlife Agencies, 266
Western North Carolina University, 224
wetland conservation, 220
wetlands, 169, 242
Whetstone Mountains (Arizona), 2
White, Aron, 232
White Mountain Apaches, 256
Wilcox, Sharon, 24, 75
Wild Cat Research and Conservation Center, 215, 238
Wildlands Network, 261
Wildlife Conservation Society (WCS)
   and Central Arizona-New Mexico Recovery Area, 255
   and *Jaguar* (Rabinowitz), 96–97
   and jaguar parts trade, 16, 20
   "Jaguars in the New Millennium" conference, 102–3
   and Kaa-Iya del Gran Chaco National Park, 124
   and Alan Rabinowitz, 141–42
   and territorial land claims., 121
   and 2030 Roadmap, 265
Wildlife Management Division (Arizona), 258
Wildlife of Mexico (Leopold), 160, 243

Wildlife Protection Act, 66
Wild Things, Wild Places (Alexander), 153
Williams, Martha, 257
Wilson, E. O., 135
Windward Road, The (Carr), 27
Wisconsin Ice Age, 40
wolves, 13, 143, 150
Wong, Wai-Ming, 190
woodpeckers, 254
World Bank, 97
World of the Jaguar, The (Perry), 25, 49
"World's Largest Tropical Wetland Has Become an Inferno, The" (Einhorn), 170
World Wildlife Fund (WWF), 55, 69–70, 163, 265
Worley, Paul, 224

Xpujil, 226

Yellowstone National Park, 122
Yucatán Peninsula, 52, 219, 225
Yucatán University, 226
Yukon Territories, 39

Zamorano University (Honduras), 127
Zapotec civilization, 52
Zeller, Kathy, 133–35, 137
Zimbabwe, 140
Zoo Biology (journal), 104